Energy Storage for Modern Power System Operations

Scrivener Publishing
100 Cummings Center, Suite 541J
Beverly, MA 01915-6106

Publishers at Scrivener
Martin Scrivener (martin@scrivenerpublishing.com)
Phillip Carmical (pcarmical@scrivenerpublishing.com)

Energy Storage for Modern Power System Operations

Edited by
Sandeep Dhundhara
and
Yajvender Pal Verma

WILEY

This edition first published 2021 by John Wiley & Sons, Inc., 111 River Street, Hoboken, NJ 07030, USA and Scrivener Publishing LLC, 100 Cummings Center, Suite 541J, Beverly, MA 01915, USA
© 2021 Scrivener Publishing LLC
For more information about Scrivener publications please visit www.scrivenerpublishing.com.

All rights reserved. No part of this publication may be reproduced, stored in a retrieval system, or transmitted, in any form or by any means, electronic, mechanical, photocopying, recording, or otherwise, except as permitted by law. Advice on how to obtain permission to reuse material from this title is available at http://www.wiley.com/go/permissions.

Wiley Global Headquarters
111 River Street, Hoboken, NJ 07030, USA

For details of our global editorial offices, customer services, and more information about Wiley products visit us at www.wiley.com.

Limit of Liability/Disclaimer of Warranty
While the publisher and authors have used their best efforts in preparing this work, they make no representations or warranties with respect to the accuracy or completeness of the contents of this work and specifically disclaim all warranties, including without limitation any implied warranties of merchantability or fitness for a particular purpose. No warranty may be created or extended by sales representatives, written sales materials, or promotional statements for this work. The fact that an organization, website, or product is referred to in this work as a citation and/or potential source of further information does not mean that the publisher and authors endorse the information or services the organization, website, or product may provide or recommendations it may make. This work is sold with the understanding that the publisher is not engaged in rendering professional services. The advice and strategies contained herein may not be suitable for your situation. You should consult with a specialist where appropriate. Neither the publisher nor authors shall be liable for any loss of profit or any other commercial damages, including but not limited to special, incidental, consequential, or other damages. Further, readers should be aware that websites listed in this work may have changed or disappeared between when this work was written and when it is read.

Library of Congress Cataloging-in-Publication Data

ISBN 9781119760337

Cover image: Wikimedia Commons
Cover design by Russell Richardson

Contents

Preface		**xiii**
1	**Introduction to Energy Storage Systems**	**1**
	Rajender Kumar Beniwal, Sandeep Dhundhara and Amarjit Kalra	
	1.1 Introduction	2
	1.1.1 Basic Components of Energy Storage Systems	5
	1.2 Types of Energy Storage Systems	5
	1.2.1 Chemical Energy Storage System	6
	1.2.2 Mechanical Energy Storage System	8
	1.2.3 Electromagnetic Energy Storage System	11
	1.2.4 Electrostatic Energy Storage System	12
	1.2.5 Electrochemical Energy Storage System	14
	1.2.6 Thermal Energy Storage System	18
	1.3 Terminology Used in ESS	19
	1.4 Applications of ESS	21
	1.5 Comparative Analysis of Cost and Technical Parameters of ESSs	23
	1.6 Analysis of Energy Storage Techniques	23
	1.7 Conclusion	28
	References	28
2	**Storage Technology Perspective in Modern Power System**	**33**
	Reinaldo Padilha França, Ana Carolina Borges Monteiro, Rangel Arthur and Yuzo Iano	
	2.1 Introduction	34
	2.2 Significance of Storage Technologies in Renewable Integration	35
	2.3 Overview of Current Developments in Electrical Energy Storage Technologies	38
	2.4 Commercial Aspects of Energy Storage Technologies	40

	2.5	Reducing the Costs of Storage Systems	41
	2.6	Energy Storage Economics – A View Through Current Scenario	42
	2.7	Implications for Researchers, Practitioners, and Policymakers	43
	2.8	Regulatory Considerations – A Need for Reform	44
	2.9	Discussion	46
	2.10	Conclusions	47
	2.11	Trends and Technological Modernizations – A Look Into What the Future Might Bring	49
		References	50
3	**Virtual Inertia Provision Through Energy Storage Technologies**		59

Shreya Mahajan and Yajvender Pal Verma

	3.1	Introduction	59
	3.2	Virtual Inertia-Based Frequency Control	61
		3.2.1 Concept of Virtual Inertia	61
		3.2.2 Virtual Inertia Emulation	62
	3.3	Impact of Low System Inertia on Power System Voltage and Operation & Control Due to Large Share of Renewables	63
	3.4	Control Methods for Inertia Emulation in RES-Based Power Systems	65
		3.4.1 Control Methods Without ESS for Frequency Control	66
		3.4.2 Control Methods with ESS for Frequency Control	67
		3.4.2.1 Battery Energy Storage Systems (BESS)	69
		3.4.2.2 Super Capacitors and Ultra-Capacitors	70
		3.4.2.3 Flywheel Energy Storage System (FESS)	70
		3.4.2.4 Hybrid Energy Storage System (HESS)	71
	3.5	Challenges	73
		References	73
4	**Energy Storage Systems for Electric Vehicles**		79

M. Nandhini Gayathri

	4.1	Introduction	79
	4.2	Energy Storage Systems for Electric Vehicle	82
	4.3	Types of Electric Vehicles	82
		4.3.1 Battery Electric Vehicle (BEV)	85
		4.3.2 Hybrid Electric Vehicle (HEV)	86
		4.3.3 Plug-In Hybrid Electric Vehicles (PHEV)	87
	4.4	Review of Energy Storage Systems for Electric Vehicle Applications	88

		4.4.1	Key Attributes of Battery Technologies	88
		4.4.2	Widely Used Battery Technologies	88
		4.4.3	Alternate Energy Storage Solutions	92
	4.5	Electric Vehicle Charging Schemes		93
	4.6	Issues and Challenges of ESSs in EV Applications		94
	4.7	Recent Advancements in the Storage Technologies of EVs		94
	4.8	Factors, Challenges and Problems in Sustainable Electric Vehicle		96
	4.9	Conclusions and Recommendations		97
		References		97

5 **Fast-Acting Electrical Energy Storage Systems for Frequency Regulation** 105
Mandeep Sharma, Sandeep Dhundhara, Yogendra Arya and Maninder Kaur

	5.1	Introduction			106
		5.1.1	Significance of Fast-Acting Electrical Energy Storage (EES) System in Frequency Regulation		106
		5.1.2	Capacitive Energy Storage (CES)		107
			5.1.2.1	Basic Configuration of CES	109
			5.1.2.2	CES Control Logic	112
		5.1.3	Superconducting Magnetic Energy Storage (SMES)		113
			5.1.3.1	Constructional and Working Details of SMES	113
			5.1.3.2	Basic Configuration of SMES	114
			5.1.3.3	SMES Block Diagram Presentation	115
			5.1.3.4	Benefits Over Other Energy Storage Methods	116
		5.1.4	Advantages of CES Over SMES		117
	5.2	Case Study to Investigate the Impact of CES and SMES in Modern Power System			118
		5.2.1	Literature Review		118
		5.2.2	Modeling of the System Under Study		121
		5.2.3	Control Approach		121
	5.3	Impact of Fast-Acting EES Systems on the Frequency Regulation Services of Modern Power Systems			124
		5.3.1	System Model-1		124
		5.3.2	System Model-2		128
	5.4	Conclusion			137

		Appendix A	137
		References	138
6	**Solid-Oxide Fuel Cell and Its Control**		**143**
	Preeti Gupta, Vivek Pahwa and Yajvender Pal Verma		

        Appendix A   137
        References   138

6 Solid-Oxide Fuel Cell and Its Control — 143
Preeti Gupta, Vivek Pahwa and Yajvender Pal Verma

    Abbreviations   144
    Symbols and Molecular Formulae   144
    Nomenclature   145
  6.1 Introduction   145
  6.2 Fuel Cells   147
      6.2.1 Different Types of Fuel Cells   148
      6.2.2 Advantages and Disadvantages   148
      6.2.3 Applications in Modern Power System   150
  6.3 Solid-Oxide Fuel Cell   150
      6.3.1 Mathematical Modeling   152
      6.3.2 Linearization   153
      6.3.3 Control Schemes for Solid-Oxide Fuel Cell
            Based Power System   155
            6.3.3.1 Constant Voltage Control   156
            6.3.3.2 Constant Fuel Utilization Control   156
  6.4 Illustration of a Case Study on Control
      of Grid-Connected SOFC   160
  6.5 Recent Trend in Fuel Cell Technologies   165
      6.5.1 Techno-Economic Comparison   166
      6.5.2 Market and Policy Barriers   168
  6.6 Summary and Future Scope   169
    Acknowledgement   170
    References   170

7 Lithium-Ion vs. Redox Flow Batteries – A Techno-Economic Comparative Analysis for Isolated Microgrid System — 177
Maninder Kaur, Sandeep Dhundhara, Sanchita Chauhan and Mandeep Sharma

  7.1 Introduction to Battery Energy Storage System   178
      7.1.1 Lithium-Ion Battery   178
      7.1.2 Redox Flow Batteries   182
  7.2 Role of Battery Energy Storage System in Microgrids   186
  7.3 Case Study to Investigate the Impact of Li-Ion
      and VRFB Energy Storage System in Microgrid System   188
      7.3.1 System Modelling   188

		7.3.2 Evaluation Criteria for a Microgrid System	191
		7.3.3 Load and Resource Assessment	191
	7.4	Results and Discussion	192
	7.5	Conclusion	194
		References	195
8	Role of Energy Storage Systems in the Micro-Grid Operation in Presence of Intermittent Renewable Energy Sources and Load Growth		199
	V V S N Murty, Ashwani Kumar and M. Nageswara Rao		
	8.1	Introduction	200
		8.1.1 Techniques and Classification of Energy Storage Technologies Used in Hybrid AC/DC Micro-Grids	201
		8.1.2 Applications and Benefits of Energy Storage Systems in the Microgrid System	202
		8.1.2.1 Applications and Benefits of BESS in Micro-Grid	203
		8.1.3 Importance of Appropriate Configuration of Energy Storage System in Micro-Grid	205
		8.1.3.1 Decentralized Control	206
		8.1.3.2 Centralized Control	206
		8.1.3.3 Coordinated Control	207
		8.1.3.4 Topology of BESS and PCS	208
		8.1.3.5 Battery Management System	208
	8.2	Concept of Micro-Grid Energy Management	209
		8.2.1 Concept of Micro-Grid	210
		8.2.2 Benefits of Micro-Grids	212
		8.2.3 Overview of MGEM	213
	8.3	Modelling of Renewable Energy Sources and Battery Storage System	214
	8.4	Uncertainty of Load Demand and Renewable Energy Sources	220
	8.5	Demand Response Programs in Micro-Grid System	221
		8.5.1 Modelling of Price Elasticity of Demand	221
		8.5.2 Load Control in Time-Based Rate DR Program	223
		8.5.3 Load Control in Incentive-Based DR Program	223
	8.6	Economic Analysis of Micro-Grid System	223
	8.7	Results and Discussions	224
		8.7.1 Dispatch Schedule Without Demand Response	224

		8.7.2 Dispatch Schedule with Demand Response	225

        8.7.2 Dispatch Schedule with Demand Response   225
        8.7.3 Micro-Grid Resiliency   229
        8.7.4 BESS for Emergency DG Replacement   235
   8.8 Conclusions   237
      List of Symbols and Indices   238
      References   240

9 Role of Energy Storage System in Integration of Renewable Energy Technologies in Active Distribution Network   **243**
Vijay Babu Pamshetti and Shiv Pujan Singh

      Nomenclature   244
   9.1 Introduction   246
        9.1.1 Background   246
        9.1.2 Motivation and Aim   248
        9.1.3 Related Work   249
        9.1.4 Main Contributions   253
   9.2 Active Distribution Network   253
   9.3 Uncertainties Modelling of Renewable Energy Sources and Load   254
        9.3.1 Uncertainty of Photovoltaic (PV) Power Generation   254
        9.3.2 Uncertainty of Wind Power Generation   255
        9.3.3 Voltage Dependent Load Modelling (VDLM)   256
        9.3.4 Proposed Stochastic Variable Module for Uncertainties Modelling   256
        9.3.5 Modelling of Energy Storage System   258
        9.3.6 Basic Concept of Conservation Voltage Reduction   259
        9.3.7 Framework of Proposed Two-Stage Coordinated Optimization Model   259
        9.3.8 Proposed Problem Formulation   260
            9.3.8.1 Investments Constraints   262
            9.3.8.2 Operational Constraints   262
        9.3.9 Proposed Solution Methodology   263
        9.3.10 Simulation Results and Discussions   265
            9.3.10.1 Simulation Platform   265
            9.3.10.2 Data and Assumptions   265
            9.3.10.3 Numerical Results and Discussions   266
            9.3.10.4 Effect of Voltage Profile   268
            9.3.10.5 Effect of Energy Losses and Consumption   268

| | | 9.3.10.6 | Effect of Energy Not Served and Carbon Emissions | 272 |
| | | 9.3.10.7 | Performance of Proposed Hybrid Optimization Solver | 272 |

      9.3.11 Conclusion   274
               References   275

10 Inclusion of Energy Storage System with Renewable Energy Resources in Distribution Networks   **281**
Rayees Ahmad Thokar, Vipin Chandra Pandey, Nikhil Gupta, K. R. Niazi, Anil Swarnkar, Pradeep Singh and N. K. Meena

  10.1 Introduction   282
  10.2 Optimal Allocation of ESSs in Modern Distribution Networks   284
        10.2.1 ESS Allocation (Siting and Sizing)   285
        10.2.2 ESS Allocation Methods   286
  10.3 Applications of ESS in Modern Distribution Networks   290
        10.3.1 ESS Applications at the Generation and Distribution Side   293
        10.3.2 ESS Applications at the End-Consumer Side   293
  10.4 Different Types of ESS Technologies Employed for Sustainable Operation of Power Networks   294
  10.5 Case Study   301
        10.5.1 Proposed Two-Layer Optimization Framework and Problem Formulation   302
              10.5.1.1 Upper-Layer Optimization   303
              10.5.1.2 Internal-Layer Optimization   304
              10.5.1.3 Problem Constraints   305
              10.5.1.4 Proposed Management Strategies for BESS Deployment   307
        10.5.2 Results and Discussions   308
        10.5.3 Conclusions   316
  10.6 Future Research and Recommendations   317
        Appendix A   318
        Acknowledgement   319
        References   319

Index   **329**

Preface

The power sector worldwide has undergone a huge transformation and the focus is to make it sustainable, environmental friendly, reliable, and highly efficient. As a result, a significant share of highly intermittent but clean renewable sources are being integrated into the power system using advanced technological components. The stochasticity of these renewable sources poses a big challenge to the efficient operation of the power system and storage will play a big role in its smooth and reliable operation. Technological developments have made it possible to use batteries and other storage systems for managing the operation of the power system. That's why we chose to come out with an edited book having some applications of the storage system in power system operation.

Our main aim is to illustrate the potential of energy storage systems in different applications of the modern power system considering recent advances and research trends in storage technologies. The book contains some useful case studies of the application of storage systems besides an introductory topic on modern storage system technologies. The book covers storage applications in demand response programs, supporting virtual inertia requirements of the power system, electric vehicle, distribution system planning and operation, microgrid operation, frequency control of the interconnected power system, and fuel cell as an alternative to the renewables. The applications of the storage have been demonstrated through simulation case studies of some standard practical systems. These areas are going to play a very important role in future smart grid operations. The research scholars, faculty members, and power industry people will find this book very useful and interesting.

In order to cover such different topics, we approached to various potential researchers and authors for the contribution of the chapters. We received more than 30 chapters but could accommodate only 10 chapters in this edition. We are thankful to all the authors who have contributed their chapters to this book. We were also fortunate enough to have a group of excellent reviewers who helped us in the review process.

We are also thankful to our mentor Prof. Ashwani Kumar Sharma from NIT Kurukshetra, Haryana, India for the guidance, support, and constant motivation during the editing of this book. As constant researches are going on in the field of storage technologies and power system, we will come up with a second part of the book covering some more areas of storage application in power system operation and control.

Sandeep Dhundhara
Yajvender Pal Verma

1
Introduction to Energy Storage Systems

Rajender Kumar Beniwal[1]*, Sandeep Dhundhara[2] and Amarjit Kalra[2]

[1]*Department of Electrical Engineering, Guru Jambheshwar University of Science & Technology, Hisar, Haryana, India*
[2]*Department of Basic Engg, College of Agricultural Engg. and Tech., CCS Haryana Agricultural University, Hisar, India*

Abstract

This chapter presents an introduction to the Energy Storage Systems (ESS) used in the present power system. Nowadays, renewable energy sources–based generating units are being integrated with the grid as they are green and clean sources of energy and also address environmental concerns. Therefore, electrical energy storage systems become one of the main components which deal with the grid instability that occurs due to the intermittent nature of these renewable energy sources. In this chapter, different types of energy storage systems reported in the literature have been presented. An effort has been made to discuss all the details such as the principle of operation, different components, and characteristics of each type of energy storage technology. Different characteristics of energy storage techniques are compared in tabular form with their pros and cons. The main objective of this chapter is to introduce the concept of storage techniques used in power systems and their needs and applications. Classification of storage systems has been presented based on short-term, medium, and long-term usage capacity.

Keywords: Energy storage system, distributed energy resources, microgrid, power system, renewable energy, storage techniques

---

Corresponding author: mail.rajendera@gmail.com

1.1 Introduction

Electrical energy is now becoming the backbone of every country as it directly helps to enhance economic growth through its several applications in the national grid. The development of any industrial infrastructure is directly associated with the availability of electrical energy and due to this growing industrial infrastructure, the electricity demand is also increasing. As per the hike in demand for electrical energy, there is a need to enhance the generation, transmission, and distribution network capacity. Initially, the basic objective of the power system network was to generate electrical power in a remote area (due to natural resource availability and pollution issues), transmit this electrical power from a remote area to the load center by high voltage transmission lines and finally distribute the power as per customers' required voltage level. After deregulation, the power system goes through so many structural changes, policy implementations, reforms, and technological advancements. The original power system is restructured in a modern power system with flexibility in operation and two-way power control. It results in a complex and huge size power system by the interconnections of different components at each level. With the increased complexity, the control of the power system becomes a tedious task. In developing countries, the demand for electrical energy is always greater than that for the generation capacity. Further, electricity demand is highly dynamic or random and difficult to predict precisely. Initially, the task of the power system was to generate, transmit, and distribute energy under central control in a vertically integrated market. But due to research and development, advancement in technology and change in regulation significantly change the framework of power systems [1]. After the reforms, the central control is distributed and a restructured modern power system comes into the existence. In this modern power system, each shareholder, that is energy producers and consumers, has active participation and has some rights. Small power producers are also encouraged to contribute and a new term "prosumer" is introduced. Prosumers are the customers who produce and consume electrical power locally. So, with this, the power flow is bidirectional and complexity is further increased. On the other hand, the operation of the power system becomes difficult due to bidirectional power flow.

Also, the majority of the share of electrical energy comes from traditional/conventional fossil fuel–based energy sources. These sources contribute to environmental pollution and greenhouse gases emission. Due to the depletion of fossil fuels, their non-availability in each country, and environmental concerns, an alternate source of energy is needed. So, nowadays, the focus on

INTRODUCTION TO ENERGY STORAGE SYSTEMS   3

renewable energy resources/sources (RERs/RES) is more which are available across the globe. With the new policies, small power producers and renewable energy generation have been encouraged to meet the energy demand. In this case, the renewable generation can be much closer to the load and named as distributed generation. It saves the transmission resources, as well as losses, which are very less. Figure 1.1 shows the concept of distributed generation (DG) in modern power systems [2].

With the focus on renewable energy sources for electricity generation, a lot of research and advancement is going in this field and this results in reduced cost and increased efficiency of these systems. This makes RERs capable to meet the small as well as large energy demands of the customers. On the other hand, if we are considering the case of solar photovoltaic (PV) and wind, which are the main resources of renewable energy generation in India, they are highly dependent on environmental conditions like solar radiation and wind speed. Due to this, the output of these sources is not continuous, i.e., intermittent in nature which limits their use. If we consider the example of PV generation, it is high in the daytime and reduces at the evening hours, which may not meet the load demand. Renewable energy resources provide

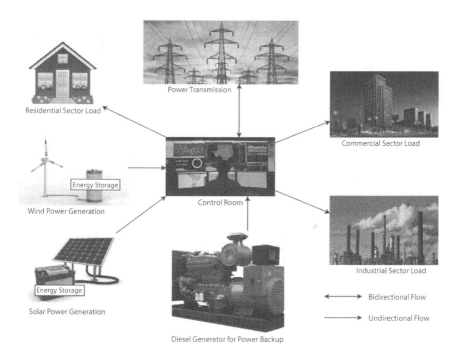

Figure 1.1 Distributed generation in power system.

a provision for clean and green energy and can be integrated with the grid or used as a stand-alone unit. But the problem associated with renewable energy resources is that these are random and uncertain, which causes issues in electricity output. So, these can't be used beyond a limit due to highly fluctuating output which causes power quality issues and makes the grid unstable and unsafe [3]. Secondly, the consumption rate of energy is not constant or fixed, it is varying with time so sometimes the availability of energy is surplus and sometimes there is a shortage; both conditions are undesirable.

Several issues like power system transients and dynamics, steady-state analysis, automatic generation control, voltage profile, active and reactive power control, spinning reserve, low voltage ride through, unit commitment and transmission curtailment, etc., are faced and for the stable and secure operation of the modern power system, these issues should be resolved. For a demand-supply balance and previously mentioned issues in modern power systems, energy storage methods are presented by the researchers.

The application of energy storage systems (ESSs) has been identified as a possible solution to compensate for the challenges faced by the modern electric power system. ESSs are in urgent demand by the conventional power generation industry, DERs, and intermittent renewable energy supply systems as they can offer required ancillary support and flexibility to them. Energy storage does not mean just the energy sources but they also provide the added benefits of improving reliability, stability, and also the quality of the power supply [4]. They are suitable for large (GW), medium (MW), or micro (kW) scale applications of the electric power system, depending on the task and requirement. They can be utilized for power demand balance, energy arbitrage, and reserve at the generation side; for investment deferral and frequency regulation at transmission level; for grid capacity support and voltage control at the distribution side; and for cost management and for peak shaving, etc., at the customer-side [5]. Thus, ESS is a necessary component and plays a key role in mitigating a wide range of operational challenges faced by the modern electric power system.

The main objective of this chapter is to give basic knowledge about energy storage systems. The reader will get familiar with different types of ESS and their applications in modern power systems whether it is the grid, microgrid or distribution generation domain, or grid integrated/stand-alone mode. The next section introduces different energy storage technologies (ESTs) including their types, benefits, and shortcomings. Several energy storage methods are available in the literature and can be used depending on the feasibility of the technique. A lot of work has been carried out by the researchers on energy storage systems and it is presented in this chapter with their pros and cons, and their characteristics in tabular form.

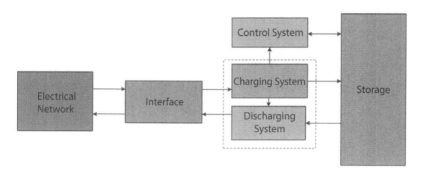

Figure 1.2 Main components of an ESS [6].

1.1.1 Basic Components of Energy Storage Systems

In AC power system, the electrical energy cannot be stored electrically. However, the AC energy can be stored by converting it into other energy forms such as kinetic, electromagnetic, electrochemical, or potential energy. Thus, a power conversion unit (PCU) is usually required by every energy storage technology to import electricity from the power grid and converting into a form that could be stored. The PCU converts it back to electricity during hours of peak demand or as when needed. Figure 1.2 demonstrates the main components of the ESS, which are as follows [6]:

i. **Storage Medium**: This provides the means to store energy for later use; such as the battery, PHES, flywheel energy storage system (FESS), capacitive/super-capacitive energy storage (CES/SCES), thermal energy storage (TES), compressed air energy storage (CAES), and superconducting magnetic energy storage (SMES).
ii. **Control**: It manages the functioning of the entire energy storage system and acts as a brain of the ESS.
iii. **Charging**: The charging unit facilitates the flow of energy from the electrical system to the energy storage medium.
iv. **Discharging**: It allows the flow of stored energy from the storage medium to the load when required.

1.2 Types of Energy Storage Systems

The issues and opportunities related to the RES encourage the researchers to provide reliable and efficient solutions in the form of ESS that have the capability to store and deliver energy when required. In addition to this,

ESS helps in maintaining power quality, load demand, grid stability, loss reduction, voltage and frequency regulation, energy efficiency improvement, reduction in fossil fuel usage, and protecting the environment from greenhouse gas emission and global warming [7, 8].

Electrical energy is not stored in its original form, but it can be converted to other forms of energy like chemical, mechanical, electrochemical, or thermal energy and can be stored (except electrostatic and electromagnetic storage, which are considered as electrical ESS) [9]. Some of the energy conversion techniques are as follows [10]:

a. Stored in the form of gravitational potential energy as a water reservoir.
b. In the form of compressed air
c. In the form of electrochemical energy using batteries
d. In the form of chemical energy by using fuel cells
e. In the form of kinetic energy using the flywheel
f. In the form of magnetic field using inductors
g. In the form of an electric field using a capacitor
h. In the form of thermal energy using sensible, latent, and chemical heat

In energy storage systems, energy is stored when it is surplus or it is in excess as compared to load and this stored form of energy is converted into electrical energy when there is a lack of energy using power conversion systems (PCS). Figure 1.3 shows different types of energy storage techniques that are reported in the literature.

1.2.1 Chemical Energy Storage System

In a chemical energy storage system, energy stored in the chemical bonds of a compound is used. In every chemical reaction, energy is released or absorbed by the reaction. In some reactions when chemical bonds break, a large amount of energy is released that can be used for electricity production.

Hydrogen is a good alternative at present for the reduction in fossil fuel. It is abundant in nature and the simplest element. Due to these merits, now hydrogen ESS (HESS) is a well-known method for storing chemical energy. In this technique, the process of water electrolysis is started when there is excess availability of electrical energy and produced hydrogen is stored in storage tanks. When it is required to supply the energy back to the grid or for any other application at that instant, the oxidation process of hydrogen

INTRODUCTION TO ENERGY STORAGE SYSTEMS 7

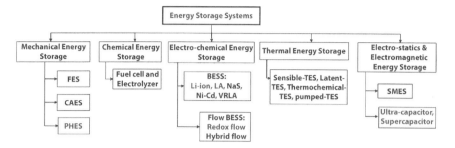

Figure 1.3 Different types of energy storage systems.

is started and electrical energy is generated from stored hydrogen. For this purpose, the setup uses a fuel cell or regenerative fuel cell to produce electrical energy from hydrogen [11] as depicted in Figure 1.4.

The chemical reaction which takes place during this conversion is [12]

$$2H_2 + O_2 \rightarrow 2H_2O + \text{Energy Produce}$$

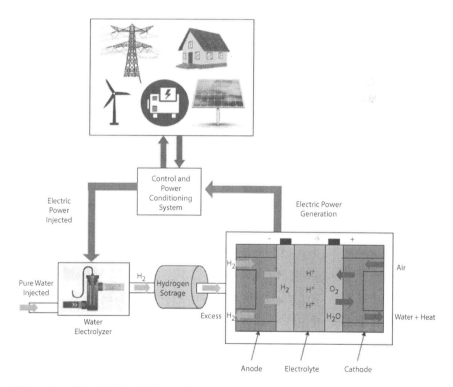

Figure 1.4 Chemical/Fuel cell or Hydrogen energy storage system.

The components of this storage system are electrolyzer, fuel cell, hydrogen storage, and power conversion system to produce electricity. The electrolyzer is the main part that is responsible for water decomposition in hydrogen and oxygen. The hydrogen produced in this process is then stored. When there is a requirement of electrical energy then fuel cells reverse this process and released energy is converted into electrical energy.

Based on the type of electrolytes or fuel used, fuel cells can be classified as polymer electrolyte membrane fuel cell (PEMFC), alkaline fuel cell (AFC), phosphoric acid fuel cell (PAFC), molten carbonate fuel cell (MCFC), solid oxide fuel cell (SOFC) and direct methanol fuel cell (DMFC) [12, 13]. The advantage of HESS is no pollution, depth of discharge (DoD) up to 100%, no or very less self-discharge rate (depends upon the type of storage). Low efficiency (40-50%), high setup cost, and low security are the disadvantages of HESS [14].

1.2.2 Mechanical Energy Storage System

It is a commonly used, effective, and sustainable solution for energy storage. In a mechanical energy storage system (MESS), electrical energy is stored in mechanical form like potential energy, kinetic energy, or in the form of pressure. In MESS, three types of storage need to be discussed: pumped hydro storage, flywheels, and compressed air technique.

a. Pumped Hydroelectric Energy Storage
In pumped hydroelectric storage power plants, gravitational potential energy of water is used. In this storage system, there are two water reservoirs which are situated at different heights [15]. When there is excess energy generation than demand at that time the water pump is used to charge or store the water from the lower reservoir to the upper reservoir. In other case, when electrical energy is required then the discharging process is used. During discharging, water is released from the upper reservoir and it strikes turbine blades and turbine feed electrical energy into the system [16]. The schematic of the pumped hydroelectric storage system is shown in Figure 1.5.

The energy (E) production equation is

$$E = mgh$$

where m is the mass of water in kilogram (kg), g is the gravitational force and h is the head or the difference in height of two reservoirs in meter.

Pumped hydro storage technologies are a very old method and a very popular one as it is used in large-scale storage.

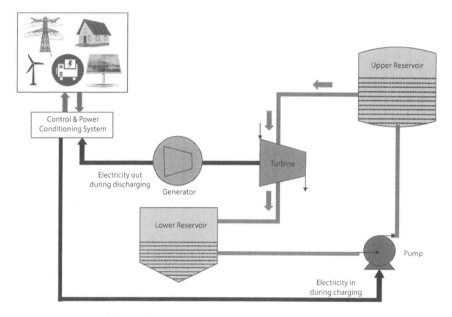

Figure 1.5 Pumped hydroelectric storage system.

As an old and matured technology, it has advantages like its per-unit cost is low, large storage capacity, the life of the system is long, and its quick response time. But there are some drawbacks of this technology, such as high initial cost, large setup time, and highly dependence on geographical area.

b. Compressed Air Energy Storage

CAES is also used for large-scale energy storage systems but these are not much wider. This storage system is similar to the gas turbine system as shown in Figure 1.6. In this storage system, the air is compressed and stored at high pressure in underground storage tanks. In the charging case, the compressor is driven by the motor, and air is stored. While in the discharging case, stored air is heated and released to the high-pressure turbine which is coupled with a generator and produces electrical power [17].

There are different types of CAES like advance adiabatic CAES, liquid air energy storage, supercritical CAES, Isothermal CAES, under-water CAES [18]. The merits of CAES are large-scale storage, high discharge rate and fast response, and while it causes pollution due to combustion and efficiency, that is highly dependent on the type of technique.

c. Flywheel Energy Storage

The flywheel is another mechanism to store the energy in mechanical form. It stores energy in the form of the angular momentum of a heavy rotating

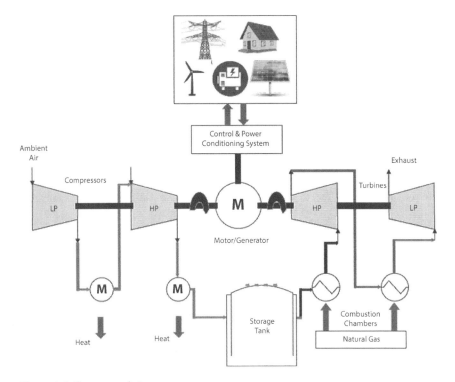

Figure 1.6 Compressed air storage system.

disk [19]. This technology is very much matured and old, as it is used in internal combustion engines. In the charging cycle, the flywheels store the energy in heavy mass by increasing the rotation speed through a motor. The stored energy depends upon the speed of the flywheel. When electrical energy is required then the discharging process starts. The stored mechanical energy is converted into electrical energy using a generator. The main components in grid-integrated mode of the flywheel energy storage system are shown in Figure 1.7.

The stored energy (E) is given as

$$E = \frac{1}{2}J\omega^2$$

where J is moment of inertia and ω is the angular velocity.

$$J = mr^2$$

where m is the mass and r is the radius of the flywheel.

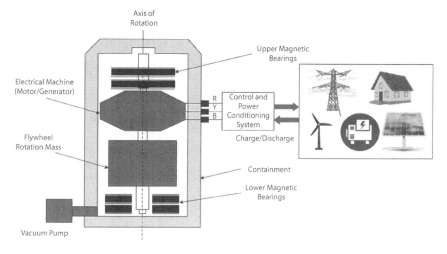

Figure 1.7 Flywheel storage system.

$$E = \frac{1}{2}mr^2\omega^2$$

In mechanical systems, friction reduces the performance of the systems. So, to increase the performance and reduce the friction, magnetic bearings are used and the whole arrangement is enclosed in a vacuum chamber to reduce aerodynamic effect [20]. Based on rotation speed FES can be classified as low-speed FES whose speed is less than 10000 rpm and high-speed FES whose speed is from 10^4 to 10^5 rpm.

Some of the key features of FESS are very fast response, quickly supply a large amount of power, long life cycle, and no impact on the environment. The self-discharge rate and the very large cost are the barriers to this technology.

1.2.3 Electromagnetic Energy Storage System

In this type of storage system as shown in Figure 1.8, electrical energy is stored in the form of a magnetic field, which is created by passing a DC current to the large superconducting coil (niobium-titanium material). As superconductor phenomena exist at low temperatures, so here coil is cooled cryogenically at a lower temperature around -270°C. In the charging state magnitude of the current is increased while it is decreased in the discharging state. PCS controls the output of electrical energy as per the demand of the power system [21].

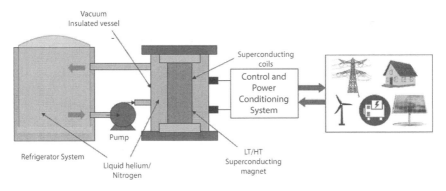

Figure 1.8 Electromagnetic energy storage system.

The stored energy in this storage system depends on the current and inductance of the coil and given as energy stored in the simple inductor as described by equation given below:

$$E = \frac{1}{2}LI^2$$

The electromagnetic energy storage system can be classified into two categories on the basis of temperature range [22]. First is low-temperature superconductor which is in the temperature range of -268°C or below. The second is high-temperature superconductors with a temperature range of higher than -268°C. This type of storage has advantages of very high efficiency, high power capacity, and long life cycle while high cost and effects due to strong magnetic field are issues associated with it.

1.2.4 Electrostatic Energy Storage System

In electrostatic ESS, the storage of energy is done in the form of an electric field. For this purpose special type of capacitor which is called a double-layer capacitor (DLC) or ultra-capacitor, is used [23] as illustrated in Figure 1.9. In this special type of capacitor design, two separate electrodes are used which are porous and a separator is inserted in between these electrodes for storage of energy in the form of the electrostatic form [24].

The energy storage capacity is given by the simple capacitor equation as given below:

$$E = \frac{1}{2}CV^2$$

INTRODUCTION TO ENERGY STORAGE SYSTEMS 13

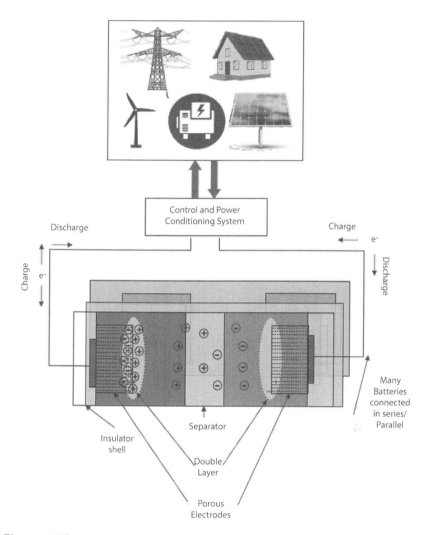

Figure 1.9 Electrostatic/supercapacitor energy storage system.

As it is evident from the equation, storage energy depends on the capacitance of the capacitor and the square of the voltage across it.

The very fast response time in milli-seconds, very long life span, and high efficiency are the key features of this technology. Operating temperature limits, high cost, and low energy density are the challenges.

There is no chemical reaction that takes place in the electrostatic and electromagnetic techniques. Further in both these storage systems energies are stored in electrical form, so these two techniques are also known as an electrical energy storage system.

1.2.5 Electrochemical Energy Storage System

Electrochemical energy storage is the most popular form of energy storage nowadays. In this system, there is a conversion of chemical energy to electrical energy and vice versa. Batteries are of two types, the first one is primary batteries and the other is secondary batteries. Primary batteries can't be reused, while secondary batteries can be reused after charging. Electrochemical energy storage batteries are secondary storage devices [25]. A generalized diagram for the basic structure of the battery is illustrated in Figure 1.10.

A battery consists of multiple cells and each cell consist of two electrodes, i.e., anode and cathode, submerged in an electrolyte. When an electrical load is connected to these electrodes externally flow of electron takes

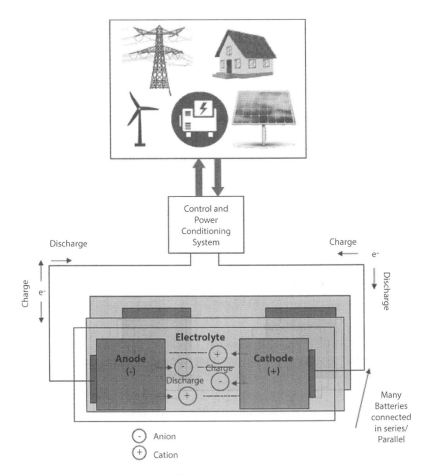

Figure 1.10 Generalized diagram of batteries.

place in the closed path from one electrode to another electrode through load and electrolyte. The terminal voltage of each cell is near about 2 V and these are connected in series/parallel configuration to get the desired voltage and current level. The batteries can be further categorized based on electrolyte and electrodes and are discussed below.

a. Lead-acid battery
Lead-acid battery (LAB) is cheaper technology and mostly uses secondary battery which was invented in 1859. LAB consists of two electrodes and electrolytes namely Pb, PbO_2, and H_2SO_4. In LAB chemical reaction takes place while charging and during discharging [26].

$$Pb + PbO_2 + 2H_2SO_4 \leftrightarrow 2PbSO_4 + 2H_2O$$

The operating temperature range is -5°C and 40°C. The drawback of the LAB is its longer charging time and its size and weight as compared to other batteries and also the life cycle of the LAB is short. In spite of all these drawbacks LAB is frequently used for uninterrupted power supply application.

b. Nickel-cadmium battery
The positive electrode of this battery is made up of nickel hydroxide or nickel oxide hydroxide and a negative electrode is cadmium or cadmium hydroxide, while potassium hydroxide is used as electrolyte [2, 8]. The terminal voltage is about 1.2 volt. The chemical reaction takes place inside the battery

$$2NiOOH + Cd + 2H_2O \leftrightarrow 2Ni(OH)_2 + Cd(OH)_2$$

The advantage of Ni-Cd batteries is their high-power density and high efficiency along with less internal resistance. Its charging and discharging cycle is fast. Toxicity and high cost are the major disadvantage of this storage technology.

c. Sodium-sulfur battery
This battery is made from non-poisonous and less expensive material. It operates at a higher temperature of about 300°C. Sodium-sulfur battery is different from others since it utilizes solid electrolytes and electrodes in the molten phase. Molten sodium and molten sulfur are used as an electrode. While in the discharge process electrons release from sodium metal and the sodium ion moves to the positive electrode through the electrolyte [27].

$$2Na + 4S \leftrightarrow Na_2S_4$$

The high operating temperature and pure sulfur are risky since it can catch fire when it comes in contact with air or moisture.

d. Lithium-ion (Li-ion) battery

This battery is the most commonly used in low power consumption devices such as portable electronic devices or mobile phones. In a Li-ion battery cathode is made up of lithium metal or lithium compound. The anode is made up of graphite carbon material and the electrolyte used is lithium salt. The terminal voltage of the Li-ion battery is about 3.6V [28].

It has energy density and power density of 100-200 Wh/kg, 1000-2000 W/kg, respectively. The advantage of this battery is its portability, lightweight, economical, and fast charging/discharging cycle. Despite this, the major disadvantage of the battery is its poor heat handling capability.

e. Flow battery energy storage system

These batteries are different from previously discussed secondary batteries as depicted in Figure 1.11. In this battery, two soluble electrolytes are stored in two different separable tanks which convert chemical energy to electrical energy by the electrochemical cell. The energy density of the flow battery depends on the size of the electrolyte tank whereas power density is based on the design of the cell. Flow battery can be classified as redox-oxidation (redox) flow battery and hybrid flow battery. In Redox flow battery, electroactive material is dissolved in a liquid electrolyte while in a hybrid one or more components are deposited. In a redox flow battery electroactive material is stored in separate tanks that circulate through pump [29].

Flow battery has few major advantages over other batteries as the capacity of this battery can be increased by increasing the size of the tank and the chemical and physical properties of electrode remain intact since it does not contain electroactive material due to which it gives stable and longer performance. Different kinds of flow batteries used in the power system are vanadium redox battery, zinc-bromine, and polysulfide bromine.

i. Vanadium redox battery

This type of battery uses V^{2+}/V^{3+} ions at the negative electrode and V^{4+}/V^{5+} ions at the positive electrode. In the charging process, V^{3+} ions converted in V^{2+} ion and V^{4+} ion converted in V^{5+} and energy is stored in the cell, while in case of discharge, this chemical energy gets converted into electrical energy. In the charging process, oxidation of vanadium, ion takes place and during the discharging process reduction of vanadium, ion takes place [30].

The energy density of the vanadium redox battery is 35-65Wh/kg and the power density of 75-150W/kg. The advantage of VRB is that

INTRODUCTION TO ENERGY STORAGE SYSTEMS   17

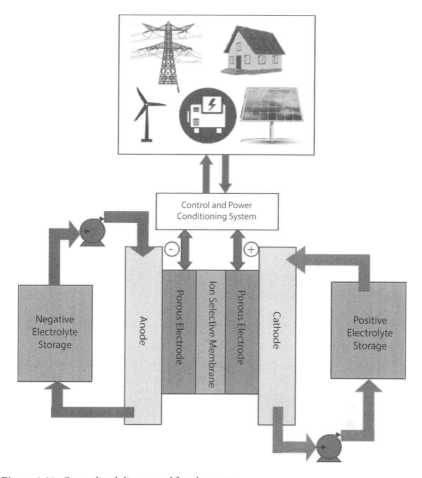

Figure 1.11 Generalized diagram of flow batteries

its efficiency is about 85%, longer life span, small self-discharge, and can be fully discharged without any damage. The standard terminal voltage of VRB batteries is 1.25V.

ii. Zinc-bromine battery
This battery comes under hybrid flow batteries. In zinc-bromine batteries [31] the tank contains an aqueous solution of zinc and bromine separately in each tank which is used during the charging and discharging process and the reversible electrochemical reaction takes place at electrolyte cell when aqueous solution flows throw it. During electrochemical reaction zinc and bromine ions move through porous separator towards opposite electrolyte direction. In the charging process, Zinc gets deposited on the anode and bromine on the cathode

while during discharging deposited zinc gets dissolved into the electrolyte through the reversible electrochemical process.

The storage capacity of these batteries is very high due to which it is available commercially for utility application. Along with all the advantages, it has several disadvantages that limits its uses such as lower efficiency and suffer from metal corrosion as compared to VRB. The mobile application is difficult due to the liquid electrolyte used in Zn-Br batteries. The new technology batteries are using fire retardant material gel in place of liquid.

iii. Polysulfide battery

Polysulfide battery is a regenerative fuel cell. In this battery sodium bromide and sodium polysulfide, the two-salt solution electrolyte undergoes a reversible electrochemical reaction. Sodium ions are allowed to transfer between electrodes through a separator made up of polymer membranes. In the charging process, three bromide ions get converted into tribromide ion through the oxidation process at the cathode and sodium particle in polyelectrolyte solution gets converted into sulfite ion at the anode. During the discharging process reverse reaction takes place [32]. Its efficiency is about 75% and its life span is about 20 years. Since the materials used in this battery are available in nature in ample amount, the cost of the battery is very low, whereas the crystal formation of bromine and sodium sulfate is the major drawback of this technology.

1.2.6 Thermal Energy Storage System

In thermal energy storage system, energy is stored in the form of ice or heat, that can be used when required. The application of this energy is suitable for industrial and domestic purposes like heating or cooling. Thermal ESS based on temperature range is classified into two categories, i.e., low-temperature thermal ESS which operates below 200°C temperature and high-temperature thermal ESS above 200°C. High-temperature thermal ESS is further classified as sensible heat, latent heat, and thermochemical system [33]. In sensible heat thermal ESS, there is a change in temperature of the material used while in the case of latent heat there is a phase change of the material that takes place. In the thermochemical system, there is a change in the chemical strength of the material by thermal induction.

A pumped thermal energy system is an example of TESS [34] as shown above in Figure 1.12. In this system compressor and turbine are connected with an electrical machine (motor/generator) for the charging and discharging process. High temperature and low-temperature chambers are used, which are connected through pipes for the exchange of heat. The chambers

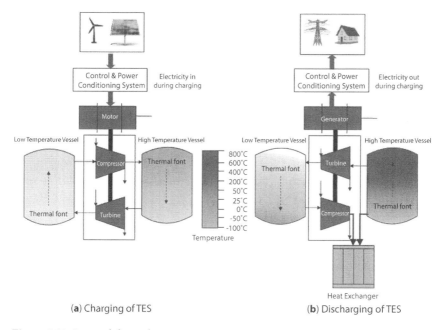

(a) Charging of TES (b) Discharging of TES

Figure 1.12 Pumped thermal energy storage.

are filled with argon gas because it heats up and cools down quickly compared to air. In the charging process, the compressor increases the temperature and pressure of argon for storage in a high-temperature chamber while in discharging mode reverse process takes place through the turbine.

Advantages of thermal energy storage are: large power storage capacity, the self-discharge rate is very less and there is no ill effect on the environment. But along with these advantages, low power density and less efficiency of conversion are the issues faced with this storage system.

1.3 Terminology Used in ESS

A number of terms used in different energy storage systems are presented in this section [2, 35]. These terms will help in understanding the characteristic of the energy system and will also be helpful in their comparison and selection of the energy storage system based on application.

i. **Power Rating**: Power rating defines how much power can be stored or delivered by the system. In other words, the power rating is the maximum output power that can be used under normal conditions.

ii. **Self-Discharge Rate**: It is defined as leakage current flowing between anode and cathode terminals even if it is not in use. Lower the self-discharge rate longer will be the discharge duration.
iii. **Discharge Rate**: It is the rate at which a battery can be discharged for the conversion of electrical energy. If the storage is discharged at a faster rate than normal, it lowers its storage efficiency and storage can also be damaged.
iv. **Depth of Discharge**: It is the percentage or fraction to which the battery can be discharged from its full capacity without any ill effects.
v. **State of Charge**: It is the available power present in the system compared to its full rating. If SOC is 100% then it means storage is fully charged.
vi. **Response Time**: The rate at which response is available concerning any application or time required to reach from fully charged to fully discharged state.
vii. **Power Density**: Power density is directly related to the size of the storage system. It is measured in Watt per kilogram (W/kg) or volumetric power density in Watt per liter (W/l).
viii. **Energy Density**: It affects the weight of the storage system. It is measured in Watt-hour per kilogram (Wh/kg) or volumetric energy density in Watt-hour per liter (Wh/l).
ix. **Efficiency**: It is the ratio of the output power to the input power of the storage system. It is calculated for specific cycles (one charge and one discharge are defined as one cycle).
x. **Technical Maturity**: It indicates that from how long a period of time the technology is in use, i.e., it is fully grown or it is a new technique that is in an initial state of implementation.
xi. **Operating Life**: It is the time for which an energy storage system produces the rated output for which it was designed. The lifetime of a system is measured in years or based on cycles of operation.
xii. **Capital Cost**: It is the main factor for each type of storage system because it demands maximum wealth for a new setup. It depends on the power (in $/kW) and energy cost (in $/kWh) per unit production.
xiii. **Operating Cost**: This cost includes the operation and maintenance costs of the storage system.

1.4 Applications of ESS

In this section, applications of energy storage systems in the electrical power system are discussed. The increasing installations of ESS in the power system maintain a balance between energy generation and consumption. Overall, the storage systems increase the reliability of the power system operations. Some of the applications where ESS play an important role are presented below [6, 21, 35].

i. Support in Generation

- *Commodity Storage*: In case, when energy demand is less at that time energy is stored and used to meet the peak demand whenever required. This provides support to the system to maintain the constant or uniform load factor.
- *Contingency Applications*: This is the reserve capacity of the system which can be used on-demand or used during fall offline.
- *Area Control*: This is the prevention of unauthorized power flow between service providers.
- *Grid Frequency*: To make grid stable and functional even if there is a large reduction or generation case by making the frequency constant between limits.
- *Black Start*: Energy needed to start up the system at its own and energies the transmission facilities and helps in synchronizing the grid.

ii. Support in Transmission and Distribution

- *System Stability*: To maintain synchronism of different components of the transmission line and prevent system collapse.
- *Grid Angular Stability*: This is done by controlling the real power to limit the power oscillations in the grid.
- *Grid Voltage Support*: This is done by providing power to the distribution grids that help in maintain the acceptable limit of the voltage by real and reactive power injection.
- *Asset Deferral*: This refers to the saving in the cost of new transmission infrastructure by making proper utilization at rated capacity.

iii. Support in End-User Energy Services

- *Energy Management*: This term is related to the management of energy. Here load scheduling is used in which load usage is scheduled as per energy pricing at the time of use. It is used for lowering the time of use prices.
- *Unbalanced Load Compensation*: This is done by using specially designed inverters by injecting or reducing the power in the individual phase load for balancing the system.
- *Power Quality Improvement*: To maintain the wave shape, magnitude, and frequency of current and voltage waveform. It helps to mitigate the power quality disturbances like sag, swell, harmonics, etc.
- *Reliability*: It is the continuity of the supply to the end customers. It is increased by the support of ESS.

iv. Support for RES Integration

- *Frequency and Synchronous Spinning Reserve*: The randomness or intermittency of RES creates imbalance between generators and results in frequency variations. This can be handled by ESS or spinning reserve at the transmission level.
- *Transmission Curtailment reduction*: In the case of wind turbines (WT), which are usually situated in a remote area, does not have good support for transmission and distribution support. In some cases, the power output of the wind turbine needs to be reduced due to congestion and this lowers the revenue of WT. In such types of situations, ESS can store energy and it can be delivered at a time of no congestion in the network.
- *Time Shifting*: The concept of "firming and shaping" is used to alter the power profiles of wind generation. This can be done by storing the energy at the time of lower demand and delivered during the time of high demand.
- *Fluctuation Suppression*: This can be done by absorbing or injecting energy during the short duration fluctuations in the RESs output.

Apart from the above-mentioned support, the ESS provides financial benefits like cost reduction, an increase in central generation capacity, revenue increase in ancillary services, reduced demand charges, etc.

1.5 Comparative Analysis of Cost and Technical Parameters of ESSs

It is very improbable that a single ESS can be suitable for all the applications of the power system. The choice of the correct storage system depends on different parameters and requirements specified by the application. The main parameters based on which the choice can be made are rated power, bridging time, efficiency, total cost, environmental conditions, footprint of the system, restrictions, space availability, etc. Tables 1.1 and 1.2 present the techno-economical characteristics of various ESS technologies. The comparative analysis of different ESSs has been shown by considering the most common assessment parameters such as power rating, power/energy density, discharge duration, efficiency, lifetime, cycle life, response time, capital cost and self-discharge rate in a day.

The PHES, CAES, and TES are the only technologies having the power rating more than 100 MW. Higher storage capacity along with low investment risk, long discharge duration high energy density, and very long life are the factors which make these technologies suitable for grid-scale storage and energy management applications. Flow batteries, large-scale battery technologies, and fuel cell have a maximum capacity of around 50 MW that are most suitable for medium-size grid applications. The power rating of FESS, small-scale batteries, supercapacitors, and SMESs are up to 1 MW. However, the factors such as quick response time, high power density, high discharge current, long cycle life, high overall efficiency make these storages most suitable for power quality, flicker mitigation, frequency regulation and other high-power short-duration applications of the power system. Similarly, various types of batteries such as LA, Li-ion, sodium-sulphur, and nickel-based battery are available for storage applications of the power system. They are the best suitable for high energy and high-power applications. A range of battery chemistries can be used for energy storage in power system applications including load following, regulation, and energy management by adding or absorbing power from the grid. LA is highly matured and the most commonly used battery system among all other battery technologies.

1.6 Analysis of Energy Storage Techniques

The main advantages and disadvantages of different energy storage technologies are summarized in Table 1.3.

Table 1.1 Cost comparison of different types of ESS [2, 21].

ESS type	Power rating (MW)	Discharge time	Response time	% Efficiency (cycle)	Capital cost Power cost $/kW	Energy cost $/kWh
PHS	100–10000	10-100 hrs	Minimum	70-75	600-2000	5-100
CAES	1-500	8-10 hrs	Minimum	40-80	400-800	2-50
FES	0-1	sec-15 min	Less than milli sec	80-99	250-350	1000-5000
Fuel Cell	0-50	sec-24 h	Less than 1sec	20-50	10000+	-
BESS	0-50	sec-h	Seconds	60-95	300-2500	10-1500
SCES	0.001-1	sec	milli sec	97+	100-300	500-1000
SMES	0.001-5	sec	milli sec	85-99	1000-10000	1000-10000
TES	0.1-300	1-24 h	sec-minutes	30-60	200-300	10-50

Table 1.2 Life cycle and environmental impact of different ESS [21].

ESS type	Energy density (Wh/kg)	Volumetric energy density (Wh/l)	Power density (W/kg)	Volumetric power density (W/l)	Life time (Years)	Life time (Cycles)	Effect on the environment
PHS	0.5-1.5		0.5-1.5		40-60		Negative impact due to reservoir, trees, and green land is destroyed
CAES	30-60	3-6		0.5-2.0	20-40		Negative due to natural gas combustion
Lead Acid	30-50	50-80	75-300	10-400	5-15	500-1000	Negative, Lead is Toxic
NiCd	50-75	60-150	150-300		10-20	2000-2500	Toxic
NaS	150-240	150-250	150-230		10-15	2500	Toxic
ZEBRA	100-120	150-180	150-200	230-300	10-14	2500+	Small
Li-ion	75-200	200-500	150-315		5-15	1000-10000+	Very Low
Fuel Cell	800-10000	500-3000	500+	500+	5-15	1000+	Negative due to fossil fuel combustion

(Continued)

Table 1.2 Life cycle and environmental impact of different ESS [21]. (Continued)

ESS type	Energy density (Wh/kg)	Volumetric energy density (Wh/l)	Power density (W/kg)	Volumetric power density (W/l)	Life time (Years)	Life time (Cycles)	Effect on the environment
Metal Air	150-3000	500-10000				100-300	Small negative effect
VRB	10-30	16-33			5-10	12000+	Negative, Toxic
ZnBr	30-50	30-60			5-10	2000+	Very Low
SMES	0.5-5	0.2-2.5	500-2000	1000-4000	20+	100000+	Negative
Flywheel	10-30	20-80	400-1500	1000-2000	15	20000+	None
Capacitor	0.05-5	2-10	100000	100000+	5	50000+	Small
Super Cap	2.5-15		500-5000	100000+	20	100000+	Small
Al-TES (Low Temp)	80-120	80-12			10-20		Small
CES (Low Temp)	150-250	120-200	10-30		20-40		Positive
HT TES	80-200	120-500			5-15		Small

Table 1.3 Advantages and disadvantages of different ESS.

ESS type	Advantages	Disadvantages
PHS	The old and matured technique, Long life, Self-discharge rate is very low, Long duration storage with adequate efficiency.	The initial cost is high, long payback period, Installation is location-dependent and negative impact on the environment.
CAES	Long life, Large storage capacity, Matured technology, and start-up time is less.	Somewhat less efficient, less adopted technique, Installation is location-dependent and negative impact on the environment.
FES	Very less maintenance required, very fast charging and discharging, Long life, Higher power density, and negligible effect on the environment.	The self-discharge rate is higher, Energy density is low, and demands extra power for energization.
Fuel Cell	Very less maintenance required Higher energy density, the Self-discharge rate is very low and negligible effect on the environment.	Efficiency is low, Cost is high, much protection is needed due to hydrogen and small impact on the environment.
BESS	Can be used in stationary as well as transportation application, Response time is in seconds, easy capacity enhancement, and portability, Flow batteries have a very long life.	Blast and toxicity are measure issues, Life depends on discharge rate and DoD, Negative impact on the environment.
SCES	The life cycle is very high, Efficiency is also high, Very fast response time.	Cost is high, Fewer applications, Energy density is less, Storage capacity is low, and has a small impact on the environment.
SMES	The life cycle is very high, Efficiency is also high, Very fast response time, No impact on the environment.	Cost is high, Energy density is less, Storage is also low, Disturbance in communication lines.
TES	Energy density is good, used in many applications, availability of storage medium, and coat is also low.	Slow response time, Energy density is less, Efficiency is low, and growing technology.

1.7 Conclusion

The prime objective of this chapter is to familiarize the reader with the energy storage systems. The analysis of various types of ESS techniques used in modern power systems has been presented. By making use of ESS, varying energy demand can be effectively managed. The penetration of renewable energy resources in the grid is increased with the support of ESS and it also helps in the stable and safe operation of the grid. Different types of energy storage techniques, the terminology associated with these, and their comparisons are also presented to give a deep insight into storage methods. The performance of energy storage systems is bounded by the materials from which they are constituted. Hence, more research is needed to understand the reaction at the atoms or molecules level and find the causes which limit their performance. Per unit cost of storage, energy is still a challenge and the barrier in wide use of many storage systems but it can be resolved by research and design activities. Furthermore, the charge and discharge cycles and rate, energy density is also needed to enhance. Overall ESS gives huge support to the modern power system operation and to meet the energy demand.

References

1. T. Strasser et al., "A Review of Architectures and Concepts for Intelligence in Future Electric Energy Systems," *IEEE Transactions on Industrial Electronics*, vol. 62, no. 4. Institute of Electrical and Electronics Engineers Inc., pp. 2424–2438, 2015, doi: 10.1109/TIE.2014.2361486.
2. O. Krishan and S. Suhag, "An updated review of energy storage systems: Classification and applications in distributed generation power systems incorporating renewable energy resources," *Int. J. Energy Res.*, vol. 43, no. 12, pp. 6171–6210, Oct. 2019, doi: 10.1002/er.4285.
3. M. K. Saini and R. K. Beniwal, "Detection and classification of power quality disturbances in wind-grid integrated system using fast time-time transform and small residual-extreme learning machine," *Int. Trans. Electr. Energy Syst.*, vol. 28, no. 4, p. e2519, Apr. 2018, doi: 10.1002/etep.2519.
4. A. Shahmohammadi, R. Sioshansi, A. J. Conejo, and S. Afsharnia, "The role of energy storage in mitigating ramping inefficiencies caused by variable renewable generation," *Energy Convers. Manag.*, vol. 162, no. June 2017, pp. 307–320, 2018, doi: 10.1016/j.enconman.2017.12.054.
5. K. Dragoon, "Energy Storage Opportunities and Challenges," *West Coast Perspect. White Pap.*, p. 61, 2014, [Online]. Available: https://www.ecofys.com/files/files/ecofys-2014-energy-storage-white-paper.pdf.

6. D. O. Akinyele and R. K. Rayudu, "Review of energy storage technologies for sustainable power networks," *Sustain. Energy Technol. Assessments*, vol. 8, pp. 74–91, 2014, doi: 10.1016/j.seta.2014.07.004.
7. S. Koohi-Kamali, V. V. Tyagi, N. A. Rahim, N. L. Panwar, and H. Mokhlis, "Emergence of energy storage technologies as the solution for reliable operation of smart power systems: A review," *Renew. Sustain. Energy Rev.*, vol. 25, pp. 135–165, 2013, doi: 10.1016/j.rser.2013.03.056.
8. F. Díaz-González, A. Sumper, O. Gomis-Bellmunt, and R. Villafáfila-Robles, "A review of energy storage technologies for wind power applications," *Renew. Sustain. Energy Rev.*, vol. 16, no. 4, pp. 2154–2171, 2012, doi: 10.1016/j.rser.2012.01.029.
9. T. Kousksou, P. Bruel, A. Jamil, T. El Rhafiki, and Y. Zeraouli, "Energy storage: Applications and challenges," *Sol. Energy Mater. Sol. Cells*, vol. 120, no. PART A, pp. 59–80, 2014, doi: 10.1016/j.solmat.2013.08.015.
10. B. Zakeri and S. Syri, "Electrical energy storage systems: A comparative life cycle cost analysis," *Renew. Sustain. Energy Rev.*, vol. 42, pp. 569–596, 2015, doi: 10.1016/j.rser.2014.10.011.
11. V. Das, S. Padmanaban, K. Venkitusamy, R. Selvamuthukumaran, F. Blaabjerg, and P. Siano, "Recent advances and challenges of fuel cell based power system architectures and control – A review," *Renewable and Sustainable Energy Reviews*, vol. 73. Elsevier Ltd, pp. 10–18, Jun. 2017, doi: 10.1016/j.rser.2017.01.148.
12. A. Kirubakaran, S. Jain, and R. K. Nema, "A review on fuel cell technologies and power electronic interface," *Renewable and Sustainable Energy Reviews*, vol. 13, no. 9. Pergamon, pp. 2430–2440, Dec. 2009, doi: 10.1016/j.rser.2009.04.004.
13. S. Mekhilef, R. Saidur, and A. Safari, "Comparative study of different fuel cell technologies," *Renewable and Sustainable Energy Reviews*, vol. 16, no. 1. Elsevier Ltd, pp. 981–989, Jan. 2012, doi: 10.1016/j.rser.2011.09.020.
14. X. Q. HUO Xianxu, WANG Jing, JIANG Ling, "Review on key technologies and applications of hydrogen energy storage system," *Energy Storage Sci. Technol.*, vol. 5, no. 2, pp. 197–203, 2016, doi: 10.3969/j.issn.2095-4239.2016.02.011.
15. T. Ma, H. Yang, and L. Lu, "Feasibility study and economic analysis of pumped hydro storage and battery storage for a renewable energy powered island," *Energy Convers. Manag.*, vol. 79, pp. 387–397, Mar. 2014, doi: 10.1016/j.enconman.2013.12.047.
16. S. Rehman, L. M. Al-Hadhrami, and M. M. Alam, "Pumped hydro energy storage system: A technological review," *Renewable and Sustainable Energy Reviews*, vol. 44. Elsevier Ltd, pp. 586–598, Apr. 2015, doi: 10.1016/j.rser.2014.12.040.
17. Y. Huang, P. Keatley, H. S. Chen, X. J. Zhang, A. Rolfe, and N. J. Hewitt, "Techno-economic study of compressed air energy storage systems for the

18. J. Wang, L. Ma, K. Lu, S. Miao, D. Wang, and J. Wang, "Current research and development trend of compressed air energy storage," *Syst. Sci. Control Eng.*, vol. 5, no. 1, pp. 434–448, Jan. 2017, doi: 10.1080/21642583.2017.1377645.
19. R. Sebastián and R. Peña Alzola, "Flywheel energy storage systems: Review and simulation for an isolated wind power system," *Renewable and Sustainable Energy Reviews*, vol. 16, no. 9. Pergamon, pp. 6803–6813, Dec. 2012, doi: 10.1016/j.rser.2012.08.008.
20. S. M. Mousavi G, F. Faraji, A. Majazi, and K. Al-Haddad, "A comprehensive review of Flywheel Energy Storage System technology," *Renewable and Sustainable Energy Reviews*, vol. 67. Elsevier Ltd, pp. 477–490, Jan. 2017, doi: 10.1016/j.rser.2016.09.060.
21. H. Chen, T. N. Cong, W. Yang, C. Tan, Y. Li, and Y. Ding, "Progress in electrical energy storage system: A critical review," *Prog. Nat. Sci.*, vol. 19, no. 3, pp. 291–312, 2009, doi: 10.1016/j.pnsc.2008.07.014.
22. V. S. Vulusala G and S. Madichetty, "Application of superconducting magnetic energy storage in electrical power and energy systems: a review," *Int. J. Energy Res.*, vol. 42, no. 2, pp. 358–368, Feb. 2018, doi: 10.1002/er.3773.
23. A. Burke, "Ultracapacitors: Why, how, and where is the technology," *J. Power Sources*, vol. 91, no. 1, pp. 37–50, Nov. 2000, doi: 10.1016/S0378-7753(00)00485-7.
24. A. González, E. Goikolea, J. A. Barrena, and R. Mysyk, "Review on supercapacitors: Technologies and materials," *Renewable and Sustainable Energy Reviews*, vol. 58. Elsevier Ltd, pp. 1189–1206, May 2016, doi: 10.1016/j.rser.2015.12.249.
25. M. Aneke and M. Wang, "Energy storage technologies and real life applications – A state of the art review," *Appl. Energy*, vol. 179, pp. 350–377, 2016, doi: 10.1016/j.apenergy.2016.06.097.
26. K. S. Ng, C. S. Moo, Y. C. Lin, and Y. C. Hsieh, "Investigation on intermittent discharging for lead-acid batteries," *J. Chinese Inst. Eng. Trans. Chinese Inst. Eng. A/Chung-kuo K. Ch'eng Hsuch K'an*, vol. 32, no. 5, pp. 639–646, 2009, doi: 10.1080/02533839.2009.9671546.
27. A. Bito, "Overview of the sodium-sulfur battery for the IEEE Stationary Battery Committee," in *2005 IEEE Power Engineering Society General Meeting*, 2005, vol. 2, pp. 1232–1235, doi: 10.1109/pes.2005.1489556.
28. J. Schnell et al., "All-solid-state lithium-ion and lithium metal batteries – paving the way to large-scale production," *J. Power Sources*, vol. 382, pp. 160–175, Apr. 2018, doi: 10.1016/j.jpowsour.2018.02.062.
29. C. He et al., "Flow batteries for microfluidic networks: Configuring an electroosmotic pump for nonterminal positions," *Anal. Chem.*, vol. 83, no. 7, pp. 2430–2433, Apr. 2011, doi: 10.1021/ac200156s.

30. E. Mena, R. López-Vizcaíno, M. Millán, P. Cañizares, J. Lobato, and M. A. Rodrigo, "Vanadium redox flow batteries for the storage of electricity produced in wind turbines," *Int. J. Energy Res.*, vol. 42, no. 2, pp. 720–730, 2018, doi: 10.1002/er.3858.
31. M. Wu, T. Zhao, R. Zhang, H. Jiang, and L. Wei, "A Zinc-Bromine Flow Battery with Improved Design of Cell Structure and Electrodes," *Energy Technol.*, vol. 6, no. 2, pp. 333–339, Feb. 2018, doi: 10.1002/ente.201700481.
32. C. Ponce de León, A. Frías-Ferrer, J. González-García, D. A. Szánto, and F. C. Walsh, "Redox flow cells for energy conversion," *J. Power Sources*, vol. 160, no. 1, pp. 716–732, 2006, doi: 10.1016/j.jpowsour.2006.02.095.
33. I. Dincer, "Thermal energy storage systems as a key technology in energy conservation," *Int. J. Energy Res.*, vol. 26, no. 7, pp. 567–588, Jun. 2002, doi: 10.1002/er.805.
34. I. Sarbu and C. Sebarchievici, "A Comprehensive Review of Thermal Energy Storage," *Sustainability*, vol. 10, no. 2, p. 191, Jan. 2018, doi: 10.3390/su10010191.
35. A. F. Zobaa, *Energy Storage – Technologies and Applications*. InTech, 2013.

2

Storage Technology Perspective in Modern Power System

Reinaldo Padilha França[1†], Ana Carolina Borges Monteiro[1*], Rangel Arthur[2] and Yuzo Iano[1]

[1]School of Electrical and Computer Engineering (FEEC), University of Campinas (UNICAMP), Campinas-SP, Brazil
[2]Faculty of Technology (FT), University of Campinas (UNICAMP), Limeira-SP, Brazil

Abstract

Energy storage is considered as the solution to the bottlenecks of intermittent generation of renewable sources like solar and wind. In many countries energy storage is being used to solve other demands of the electrical system such as eliminating overloads in the transmission lines or even stabilizing the voltage variations occurring in the distribution networks. This technological method used to store electricity is needed in communities located in remote regions, without access to the electricity grid. Thus, storage systems are essential for the functioning of electronic equipment since sometimes the distribution of electrical energy in these locations is not available. Conceptually energy storage systems mediate variable sources and loads, "moving" energy over time, that is, energy generated at a given moment can be used at another. This technology converts electrical energy into another form of storable energy, i.e., chemical, mechanical, thermal, among others. It attempts to solve a problem common to electrical energy produced through renewable primary sources, given their characteristic of producing electrical energy only according to the momentary availability with great variability. The generated power cannot be controlled precisely. Therefore, this chapter aims to provide an updated overview of Energy Storage systems, besides their application in a Modern Power System. It addresses its evolution and branch of application potential in the industry, showing and approaching its success relationship, with a concise bibliographic background and synthesizing the potential of technology.

---

*Corresponding author: monteiro@decom.fee.unicamp.br
†Corresponding author: padilha@decom.fee.unicamp.br

Sandeep Dhundhara and Yajvender Pal Verma (eds.) *Energy Storage for Modern Power System Operations*, (33–58) © 2021 Scrivener Publishing LLC

Keywords: Energy storage, power system, sustainable power networks, energy markets participation, renewable energy technologies

2.1 Introduction

Energy storage systems are the methods and technologies used to store electrical energy. These systems are needed mainly by populations living in remote regions, and therefore they do not have access to electricity. This reality is very common in underdeveloped and developing countries, since this equal distribution of resources often depends on governmental interest [1-2].

But not only the most remote regions suffer from a shortage of electricity; many urban areas also face lots of power cuts. The declines in the quality and quantity of electricity are becoming a serious challenge, due to the fact that global warming has generated a scarcity of water for hydroelectric and for resources for thermoelectric and nuclear energies [2, 3].

Therefore, the development of renewable energies such as solar, wind, and batteries has been identified as a solution. However, this solution cannot be applied globally, as not all places in the world have enough sunlight for a time and intensity enough to meet the needs of a home, an industry, or a city. The same applies to wind energy, which not only requires adequate climatic conditions, but also a large space for the installation of all machinery for power generation. The energy produced by these renewable and non-polluting sources is distributed by the network in real time [3].

Thus it is important to have storage as generating energy as solar and wind energy are intermittent. For years, industries have been using storage technologies in electronic equipment, vehicles, and turbines. Some alternatives have already been developed which include lead-acid batteries and lithium-ion batteries and contribute to the stabilization of energy supply systems [4]. In many countries, energy storage has been used to stabilize voltage variations that occur in distribution networks besides eliminating overloads in the transmission lines. Regarding Energy Storage, it is possible to carry out several applications that make the electrical system more robust and reliable, providing a quality electrical supply [5-7].

The main advantages of the energy storage system are related to (1) the economic factor, since the energy produced in periods of greater supply can be used in periods of scarcity; (2) scalability, since rapid implementation allows for the expansion of installed capacity according to the needs of consumers; (3) reliability when the transmission occurs without failures and/or interruptions; (4) and sustainability related to the reduction of environmental impacts [6, 7].

Electric energy storage systems provide greater flexibility, especially for industries, but it is also very desired by non-industrial consumers, because through storage it is possible to reduce the final cost of the service. In this sense, solar energy has been incorporated into the daily lives of ordinary people, especially those who live in tropical countries. Homes in regions with a higher incidence of sunlight have received solar panels on their roofs in order to reduce monthly electricity costs. In some places in Brazil, people with low financial income receive this technology in their homes, right at the time of purchase of these homes sold at low cost by the government [6, 7].

These commercial-scale systems have already proven to be viable as the prices of the systems falling and reduced incentives for renewable energy from energy storage are likely to become more profitable than exporting surplus energy to the grid. In other words, for storage to become viable in applications in homes, companies, and industries, it is necessary to include policies aimed at reducing the costs of storage systems and the correct selection of tariff values [8, 9].

Therefore, this chapter aims to provide an updated overview of Energy Storage, as well as showing its context on the horizon in the Modern Electrical System, addressing its evolution and application potential in the industry, addressing a concise bibliographic background, summarizing the potential of technology.

2.2 Significance of Storage Technologies in Renewable Integration

The viability of commercial-scale storage systems increases with the drop in technology prices and incentives to renewable energies being fed to the electricity grid. Access, cost, and quality are issues in rural areas where agribusiness has major consumption and affect the economy of many countries [8].

The great economic relevance of agriculture is paramount in developing countries like Brazil, given the consequence of the wealth of natural resources, availability of area, and favorable climate. Assessing that in rural areas, access to the use of the electricity grid as an economic advantage is fundamental in terms of energy storage. The supply for buildings in urban areas has different patterns in contrast to the energy supply in rural areas [9].

In view of this, the agribusiness is extremely intense in energy, deciding a significant portion of its production costs. In addition to dealing with more distant regions, rural producers often suffer from a deficiency in the electrical structure such as poor quality or even a lack of energy. In general, a diesel generator is used to supply the need of power in rural areas, but they are expensive and harmful to the environment. Therefore, photovoltaic energy stands out as the best solution in this scenario [10].

Before selecting the best type of energy to be used, it is necessary to consider in advance the relationship between the electricity generated and consumed, the cost of electricity, the tariffs imposed by the government, and the cost-benefit of renewable energy eligible to be used. For producers in rural areas that do not have a connection to the concessionaire's network, the off-grid photovoltaic system, that is, the system disconnected from the distribution network, in addition to the photovoltaic modules and inverters, a battery bank that stores the energy generated during the day for night use and/or on rainy days, it appears as a replacement option for diesel generators [10–12].

In relation to traditional generators solar power generation has many advantages, which are (1) reduced production costs, (2) the independence of the company that supplies electricity to the country in relation to on-grid systems. On the other hand, the main disadvantage is the greater capital cost as not only panels are costly but the battery banks also have a high cost of implementation, but in the long term, the investment employed is fully recovered [2–12].

The consumers have an alternate option for photovoltaic systems since they can use the traditional electricity grid as a form of storage, eliminating the need for the battery bank. This will have a lower cost of implementation and, consequently, giving faster return for the investment made [12–15]. Employing renewable energy for agricultural production brings benefits in terms of point monthly cost of energy. The use of renewable and non-polluting energies even reduce the final cost of the product to the final consumer, and also increase agricultural production, receiving quality seals from environmental and sustainability agencies of the government. Such characteristics improve the image of rural products, as many consumers today are concerned with the type of product they are consuming and what environmental impacts are generated on the consumer by determining the product of a specific rural producer [4–13, 15].

In addition, rural photovoltaic energy is an alternative to reduce these losses, improving the performance of irrigation, refrigeration, milking, storage, and processing of rural products. The intermittency in renewable energy may lead to instabilities or breakdown in the supply of the conventional electrical grid resulting in immeasurable losses to the producer and consequently to the country's economy [14, 15]. In relation to the hotel and tourism sector, energy consumption through clean and low-cost energy sources (solar and wind) results in a significant reduction in the operating costs of hotels, since they have a high energy consumption in heating water for showers, swimming pools, bathtubs, taps and even for washing bedding and towels, among others [16].

This energy (solar photovoltaic energy) is "treated" by a device called an inverter, making it possible to use it to power any electrical equipment. The use of clean and renewable energies for the operation of a hotel or tourist place can be a differential in attracting guests and tourists, since society increasingly seeks sustainable alternatives [17–19].

Analyzing solar energy more specifically, it is captured by means of solar cells that capture some wavelengths of solar radiation, that is, the solar rays when they reach the photovoltaic cells of a solar module and cause a reaction on an atomic scale, called photons (energy present in the sun's rays). When they collide with silicon electrons there is the generation of electricity and consequently an electric current [20].

Even in places with more moderate climates, or with most of the weather cloudy and covered by clouds, solar radiation can still be a source of energy. Solar electricity is one of the cleanest, most abundant, sustainable, and renewable sources in the world. Among the advantages of a solar-powered storage system, in addition to economy and sustainability, reliability is greater in the electrical system. Considering the possibility of its use in conjunction with other equipment, such as storage devices, gives greater stability to the electrical network, and can even supply part of the electrical consumption when there is a failure in the distribution network [21–23]. The electricity generated by solar energy does not cause noise pollution and carbon emission [24].

Even in the midst of so many benefits, it is important to point out that wind and solar sources are not able to supply the needed energy in the necessary volumes or at the desired time always, requiring energy storage technologies. This makes it necessary to combine these efficient systems with other heat and electricity storage technologies, resulting in a significant increase in total energy storage compared to the separate implementation. Through an infrastructure connected to the electric grid the energy industry can overcome the interruption challenges [25–27].

2.3 Overview of Current Developments in Electrical Energy Storage Technologies

Solid-state batteries are electrochemical cells that convert chemically stored energy into electrical energy. They are strongly related to technological advances and the materials used, improving the performance and capacity of modern batteries. Each cell that contains a positive terminal (cathode) and a negative terminal (anode) forms an electrolyte allowing ions to flow freely between the electrodes and terminals. This process generates an electric current, creating technologies such as electrochemical capacitors that can be charged and discharged instantly and can have almost unlimited durability. Examples of solid-state batteries are nickel-cadmium (Ni-Cd), lithium-ion (lithium-ion), and molten salt (NaS) batteries. In this context it is important to note that lithium-ion is more powerful and durable when compared to sodium batteries) [28–30].

Battery energy storage solutions are closely linked to electric car technology. Lithium, on the other hand, is considered to be the lightest of the elements, guaranteeing batteries greater power distribution capacity and greater storage. However, they still present challenges related to the high cost [28–30]. Another alternative is the frozen air technology, which uses excess electrical energy to cool the air to less than 200° C, since at this temperature the air reaches its liquid form and can be stored and transported. Consequently, when heated, it expands and moves a turbine, converting mechanical energy into electrical energy. The technology is used to store energy in times of low availability of natural resources. Another feature of this type of system is low gas emissions. In addition, frozen air can be used to move the engine's pistons and air-conditioning equipment. Air liquefaction is considered a way to store energy from intermittent renewable sources (solar and wind) [31–35].

Flow batteries are a type of rechargeable battery in which the recharge is provided by two chemical components dissolved in solution and separated by a membrane. Different classes of flow batteries have been developed, including hybrids, redox, and without membrane, considering that the fundamental difference between conventional batteries and flow cells is that the energy is stored with the electrode material in the conventional one, whereas in the case of flow it is stored in the electrolytic solution. This technology has the possibility to be recharged almost instantly with the exchange of this electrolytic solution [36, 37].

Another technology developed is that of compressed water, which uses a mechanism similar to that of frozen air, but instead of air, the technology uses water. It consists of pumping and compressing water in the depths of the earth, in the cracks in the rocks of abandoned oil and gas wells. From the moment this water is released and pressurized, it acts as a source that drives a turbine, generating electricity. This concept can be used to build and operate energy storage projects for power plants, storing energy when price and demand are low, for later use when cost and need are high [24, 38, 39].

Flywheels are equipments composed of the operation of an electric motor feeding the rotor that goes into rotation and stores the angular kinetic energy. When stored energy is needed, the rotor is slowed down and mechanical energy is converted to electrical energy. Rotations occur at high speeds in a vacuum or low-pressure container so that losses due to friction with air are minimized. This technology is capable of storing energy from intermittent renewable sources and supplying electricity to the grid continuously and at the appropriate frequency. Consequently, there is an improvement in the quality of supply. This method of power generation is generally used on ships and subways, where the braking energy of trains is stored and used to start stations [40, 41].

Gravity can also be used as a form of energy storage, as it is used in water elevation stations for reservoirs. This is because water located at a lower level is pumped into hydroelectric reservoirs, so potential energy is stored [42, 43].

Hydrogen energy storage technology is based on water electrolysis and has properties to store large amounts of energy. In this case, the element is used as an energy vector in association with solar, wind, and/or hydroelectric energy. The technical feasibility consists in the use of water poured from the plant for the production of hydrogen, which can be used in fuel cells for charging electric vehicle batteries, for example, and in auxiliary energy systems [4, 44, 45].

The technology of thermal energy storage consists of generating heat or cold to store energy, since electrical energy is used to pump heat from the cold reservoir to the hot reservoir. Thus, its performance is similar to that of a refrigerator, since to generate electricity, a heat machine is used providing work for a generator. Consequently, to recover energy, the heat pump is reversed to become a heat machine. In this way the engine removes heat from the hot tank, producing mechanical work to deliver heat to the cold tank [11, 45].

Another way to use thermal storage technology is known as cryogenic energy storage. This method consists of using electricity to cool the air until it is liquefied, and then it is stored in a tank. In the face of situations that demand energy, this liquid air is converted to a gaseous state. The heat generated in industrial processes or even by exposure to ambient air can be used as a tool in this process and be used to turn a turbine, which will generate electricity [15, 45].

Reversible hydroelectric power plants are energy storage technologies and can store large amounts of energy, operating with two reservoirs. The resulting energy is generated when the water descends from the highest to the lowest point, passing through the turbines. In situations of surplus energy in the grid, these plants operate in pumping mode, operating with an electric motor pumping water from the lower reservoir to the upper reservoir. In this way, there is availability to generate energy at times of higher consumption [46]. Or, reflect on another form of energy storage while preserving the storage potential of hydroelectric ponds through hybrid generation in association with wind farms, resulting in the production of energy only when the winds do not blow, saving water, that is, potential energetic [47–49].

2.4 Commercial Aspects of Energy Storage Technologies

One of the limiting factors of renewable energies is the storage capacity of this energy, which is one of the reasons that delay the arrival of electric vehicle technology on the market in a wide and accessible way for all populations. Based on this, it is noted that renewable energy sources are indispensable for the sustainable growth of the planet as a whole. However, the energy produced by renewable sources depends on nature, and not on human demand [50–53].

The use of electrostatic energy can be used as a source in capacitors and super capacitors or magnetic energy as a medium of superconductors. The mechanical energy in the form of kinetic energy storage is composed of a steering wheel (wheels that rotate at high speed, storing kinetic energy) that can be used for any purpose, including the generation of electrical energy. However, the amount of stored energy is relatively small when compared to that stored by other devices, such as batteries; or as potential energy through hydraulic technology and compressed air [54–56].

In relation to chemical energy, storage occurs by means of electrochemical energy in the form of conventional batteries and flow batteries, fuel cells, and metal-air batteries. Thermo-chemical energy in turn works through clean energy technologies such as hydrogen, solar, and metal heated by solar energy [45, 56].

Lithium battery storage has flexible properties, and due to its high range of applications, this technique has been extensively studied and improved in recent years. Thus, its applicability varies from stationary applications, such as Smart Grids, to non-stationary applications, such as use in portable devices and electric vehicles [57, 58]. Compressed Air Energy Storage (CAES) technology is based on the generation of electrical energy by conventional gas turbines. In this process, the compression cycles are uncoupled and a turbine expands, separating into two processes. Consequently, there is the storage of electrical energy in the form of compressed air elastic potential energy. This type of technology has great commercial and industrial use as it is capable of being replicated on a large scale and meeting great energy demands [59–61].

Sodium-sulfur batteries (NAS) are a commercial energy storage technology being applied to support the electrical networks provided by companies. This methodology is still classified in the electrolytic category and has great potential in the integration of renewable energies such as wind and solar [29, 62].

Although there are different energy storage systems, the challenge is to keep the system in a robust, reliable, and economically competitive state. It is always important to consider the individual characteristics of each methodology, the energy demand, the installation environment, the rural or urban location, and the energy conversion velocity. All of these features are important as they are directly related to system performance and storage capacity [63–66].

Therefore, there are many ways to store energy, but in many cases, the type of storage device used depends on the intended application. Regardless of the application, all energy storage devices and systems depend on efficient electrochemical mechanisms that prevent electron and ions migration, but at the same time, they must ensure efficient storage [67].

2.5 Reducing the Costs of Storage Systems

The exponential increase in battery installations in electrical systems (especially lithium-ion batteries) is expected to reduce costs in the coming decades, due to their wide global diffusion related to electric vehicles. Its main benefits are seen in relation to the high energy density, low response time in operation, high efficiency, low water consumption, absence of emission of pollutants in the operation, high power in a short period of time, and greater durability [68].

Electrochemical batteries are the main technologies developed in recent years, such as lead-acid, lithium-ion, sodium-sulfur, and flow batteries. As each technology has different characteristics, technical and economic comparisons must be made on a case-by-case basis. However, a large part of the research in this sector is on lithium-ion technology, as it has predominance in new installations of network-connected storage systems [69].

Regardless of considering only battery cells or the complete energy storage system, the prospect of cost reduction is similar, since the increase in the number of installations tends to reduce the costs of other components and services associated with the system as a whole. In this sense, they can be used through the interconnected electrical network in the distributed generation of energy and isolated systems (not connected to the network). Consequently, reliable, safe, and low-cost solutions are developed. Another factor that drives the market for storage of stationary batteries is the reuse of electric vehicle batteries, as it contributes to reducing the costs of new installations [70].

Based on the reality of the expense and popularization of renewable energy sources as well as the storage methodologies, soon, new contracts, standards, and protocols should be created in the energy sector, where new rules and costs will arise linked to the distribution of these methodologies applicable to daily lives of large companies, businesses, and homes [1, 3, 5, 70].

2.6 Energy Storage Economics – A View Through Current Scenario

There are economic opportunities in the electrical network management market that have made energy storage attractive in terms of performance and cost. This is due to the possibility of using batteries attached to a photovoltaic plant, which consequently generates the model's economic viability for a system on the market. In addition, depending on the system's location, technical losses are avoided [18, 54, 71].

The technology of Net Metering is an electric energy compensation system, which acts by means of absorbed and unused energy, whether from photovoltaic solar panels or even small wind turbines. All of this is transformed into credits for discounts on the consumption of electricity in a residence, company, or industry. Net Metering deployed in homes can make consumers more autonomous to produce and distribute the energy produced by their own photovoltaic system [72, 73].

Alternatives to energy storage resources allow consumers autonomy and have less impact on the environment; after all, the energy source is clean and inexhaustible. In addition, these technologies can be used as a back-up supply, acting with automatic supply in case of failure of the generating unit and conventional distribution. In addition, it can also level the daily load curve, reduce peak shaving demands, among others. All these characteristics make the combination of clean and renewable energies with energy storage techniques the most competitive in terms of sales and implementation [74].

The development and implementation of all these technologies are in line with the Paris Agreement, whose main objective is to strengthen the global response to the threat of climate change and strengthen the countries' capacity to deal with the impacts arising from these changes. In addition, these new methodologies create new opportunities for fostering economic development in line with the Sustainable Development Goals (SDGs), composing a global agenda for the construction and implementation of public policies that address several fundamental themes for human development [75, 76].

2.7 Implications for Researchers, Practitioners, and Policymakers

Storage technologies are classified into six categories: (1) flow batteries, (2) solid-state batteries, (3) compressed air, (4) flywheels, (5) thermal pumping, and (6) hydraulic. In this case, solar and wind energy are the least polluting, but still need time for the effective development of economic viability to also determine efficiency gains in the energy generation process [77]. In this sense, the researchers are looking for new ways to store solar energy in its "pure" chemical form, that is, sunlight is transformed into electrical energy or just heat before being stored. The advantages are that light has properties that make it capable of being stored for long periods without loss of energy. In addition, only a catalyst is needed, a small temperature change leads to energy release. Fulvalene, diruthenium, is a chemical compound used to store sunlight. This substance is derived from a mixture of carbon nanotubes and the chemical compound azobenzene, which together create a new material with the most efficient and economical solar storage capacity [78–81].

There are other unconventional technologies in the research and development phase, such as Advanced Railway Energy Storage (ARES), which aims to guarantee the storage of renewable energy through electric locomotives; the result is a hybrid storage model. However, many types of research are still focused on lithium-ion, as this technology has insertion designed for electric vehicles, still allowing the use of this equipment in the worldwide transport matrix and use in electronic equipment. On the other hand, it is important to emphasize that this path leads to the growing need for batteries, which can intensify a more intense and frequent extraction of mineral resources [78–81].

The mining activity is largely responsible for the current configuration of the current technological society, since several end products and resources used come from this activity. However, it should be noted that in countries where these activities are carried out, there must be environmental licensing procedures that establish mitigation, prevention, and compensation measures that minimize the effects of negative impacts. All of this has the objective of not creating a paradigm contrary to the storage of renewable energies in the electrical matrix, thus being in total opposition to the negative environmental impact. In short, any new technology created to meet energy needs must undergo strict environmental checks. In this context, to minimize the problems related to mining, some researches seek the use of alternative and more efficient materials, given that some of these sources are recyclable materials [78–81].

2.8 Regulatory Considerations – A Need for Reform

It is increasingly clear that solar generation and other renewable energy sources are constantly growing and expanding around the world, especially in countries geographically favored for this purpose, such as Brazil. The regulation of energy storage systems is already a topic of debate in countries where the implementation of this technology is ongoing and expanding. Due to the multiple uses of this technology, it needs a specific regulatory discussion; however, regulatory aspects cannot be limited to the insertion of intermittent sources [82–85].

Those aspects that allow the contracting and valuation of specific attributes, such as the generating and transmitting agent, as well as the creation of the ancillary services market, need to be weighed. These are essential technical requirements for the Electric System to operate with quality and safety. Therefore, they are those that complement the main services, characterized by generation, transmission, distribution, and commercialization [82, 83].

A strategy for development in other countries over the years is how to hire solutions to specific problems. In view of the relevance and the opportunity of this type of contracting of the implantation, it must be be analyzed from the systemic point of view. Thus, regional benefits are considered for the service of the entire energy chain, generating reliability in the energy network and, consequently, reducing transmission needs and costs [82–85].

In Latin America, developing countries are still studying the best way to insert storage systems in their energy matrix, discussing the insertion of batteries to assist in the regulation of the primary frequency. These countries still develop standards that allow the installation of storage systems to mitigate problems in transmission and distribution. In addition, the main socio-economic and environmental aspects related to batteries must be given priority in order to consider their life cycle, production, use, recycling, reuse, and even final disposal of these products [82–85].

The technology of energy storage by means of reversible hydraulic plants (UHR) is the most widespread internationally; the regulatory and commercial aspects are focused on this technology. This technology aims to balance the frequent fluctuations between the lack and excess of electricity, capable of restoring the supply of energy to the grid without external sources of energy. Unlike thermoelectric plants, reversible hydroelectric plants can react to variations in the grid in a shorter period, producing the necessary electrical energy or even consuming the surplus [82–85].

For specific regulatory purposes of the UHR project proposal, factors associated with wind or photovoltaic generation, or even thermal generation at the base, or combinations of alternatives should be considered. Thus, it is important to value the benefits of this technology in relation to the contribution to frequency/voltage correction, quick dispatch or adjustments to load variations, and even available to the consumer, even in a scenario of hydrological scarcity [84, 85].

The principle of operation of this technology is based on the special characteristics used as energy accumulators in hydroelectric plants of a single enterprise, that is, in view of the surplus energy in the network, it switches to pumped mode, that is, an electric motor starts pump turbines, which pump water from a lower reservoir to an upper reservoir. Otherwise, in the case of an increase in the demand for energy in the network, the water from the upper reservoir is discharged to the lower reservoir through a forced duct. Consequently, the water then causes the rotation of the pump turbines, which start operating in the turbine mode, activating the generators [82–85].

As the other traditional technologies increase participation in electrical systems, they will also be analyzed necessarily, so that they do not impede the expansion of technological advances related to the development of the renewable energy and energy storage industry. Thus, through governmental stimuli, proposals for the generation of hybrid energy for consumers can be initiated [86–88].

Among the energy storage technologies, the steering wheels act by absorbing and storing kinetic energy when accelerated and return energy to the system when needed. Consequently, there is a reduction in its speed of rotation, which is used to smoothen the variations in the speed of an axis caused by the variation of the torque. The SMES (Superconducting Magnetic Energy Storage) that stores magnetic energy in a superconducting coil, the current can circulate indefinitely without the coil (short circuit) being connected to source power. This is because superconductors experience certain losses in an alternating current; or other technologies based on super capacitors and hydrogen [4, 45, 88].

The storage of energy by Hydrogen is related to the concept of generating electricity from solar energy with the production of hydrogen from superheated water, obtaining energy production 24 hours a day, seven days a week. Hydricity is a potential solution for continuous and efficient energy generation, providing an opportunity to create a sustainable economy to meet all human needs, including food, chemicals, transportation, heating, and electricity [45, 88].

2.9 Discussion

Electricity distribution requires a constant balancing act in adjusting supply to meet society's demand. From a balance point of view, if this becomes unstable, it can cause fluctuations in the energy frequency that can lead to power outages. Energy storage systems provide a wide variety of technological approaches that allow the management of energy supply, while allowing the creation of energy infrastructure with more robust and resistant characteristics. The generation of electricity from renewable energy sources (sun, wind, and the like) can still be considered which is fully dependent on climatic conditions and the speed of available resources and natural phenomena, which fluctuate easily.

At the same time as assessing the existence of several energy storage technologies, such as compressed air, superconductors, hydraulic pumping, flywheels, types of batteries, and hydrogen, among several others that incorporate large amounts of renewable energy in the network, the ways in which the conditions under which they are used affect their performance, cost, and useful life must be addressed effectively. From a technological point of view, battery storage can currently be considered established, but it does not mean that the development of this technology, in its various instances, is stagnant.

Even so, energy storage systems with frequency regulation capability, together with control systems using information technologies, i.e., computer control systems, are promising solutions that maintain a stable energy supply. Still it is necessary to evaluate the monitoring, management, and control system managing the storage system, ensuring safe use and maximizing its performance, preventing individual cells from overloading, for example, as well as acting in the control of battery charge and discharge, which generates economy and efficiency.

It is worth mentioning the existence of increasing demand for stabilization of the electricity grid as the amount of renewable energy grows. In view of the demand nowadays, this requires more than energy storage management systems. Considering that battery technology is still considerably expensive and complex, aiming at solutions in more robust systems, such as software and tools that allow remote monitoring and control for these systems, is necessary. In small-scale electrical networks and micro-networks an interaction requires harmony in need of energy storage solutions that balance supply and demand, besides its use during times of peak or shortage. In this sense, it is worth reflecting that there is no single type of technology that can be used for a single particular application, so there are many options which depend on the decision criteria and the application context, such as space limitations, performance requirements, efficiency, conditions environmental (in terms of renewable energy use), installation infrastructure, security, maintenance requirements, and costs, availability and cost of components, technology life cycle, network or utility requirements, manufacturer and supplier reliability, or even the technology life cycle.

Assessing the variety of technologies, operating principles, and materials that compose them, it is important to reflect the contribution to stable energy supplies, which are essential for people's comfort, through innovative solutions, obtained through research and development until systems integration, in addition to manufacturing materials for energy storage devices for industrial, home or urban applications.

Systems that configure economical solutions for specific applications optimize the best energy storage system for a certain application. Considering innovative technologies that include storage by lead-acid batteries, which store relatively high amounts of energy, lithium-ion batteries are the most used as they are able to provide high energy levels for a short period of time.

2.10 Conclusions

The storage of energy in electrical systems is a real trend, which for a long time has been considered a critical technology. However, it is becoming increasingly viable considering the advent of electrical networks technologies. These technologies can be used intensively in order to exploit conventional or renewable energy resources, thus playing an important role in the unification, distribution, and expansion of the capacity of distributed generation systems.

Through energy storage systems it is possible to effectively disseminate the use of renewable energies, with regard to solar, wind, and other renewables in a sustainable manner. These also relieve congestion in the electricity network, bypassing the problems of availability and randomness of power. This increases the efficiency of the system as a whole, contributing to the reduction of environmental impacts, concerning the possibility of effective use of renewable energies.

The increasing insertion of renewable energies in the energy matrixes drives the storage technologies, as it is not possible to guarantee the availability of the amount of wind or sun exactly sufficient in each hour of the year day or even control the energy injected at each instant of generation. Faced with this scenario, the batteries correspond to the opportunity of a predictable energy dispatch, since they will be able to store the generated energy. The operation of these electrical systems equipped with this type of devices allows the battery to be charged when there is no need for energy. When there is need of energy by the system, and there is not enough energy source (wind or sun) available, the battery will start operating, bringing energy flexibility to the system.

The hourly granularity is essential for the battery, as a form of energy storage, since they are the best current options, due to simplicity of its assembly, and planning in advance ensures energy adequacy. The other energy storage platforms complement renewable energy, thereby improving the performance of the network, and consequently reducing its costs.

The decline in the quality of electricity is becoming a serious challenge, as large amounts of renewable energy are introduced into the electricity grid; this ensures that homes and businesses can be powered by green energy, even when the source (sun or wind) is not present. This corresponds to the property of good quality electricity that maintains voltage and frequency at a certain level, which is essential for various installations or equipment, that is, the demand for energy storage systems to overcome the energy challenge.

Finally, energy storage systems solve a problem common to electrical energy produced through renewable primary sources, given their characteristics of producing electrical energy only according to the momentary availability having great variability, i.e., whose generated power cannot be precisely controlled.

In summary, energy storage technologies must have a long storage time, low capital cost, and particularly a relatively high efficiency compared to other current technologies, which makes it possible to be deployed.

2.11 Trends and Technological Modernizations – A Look Into What the Future Might Bring

The generation of energy by conventional methods is becoming increasingly expensive and inefficient, and in view of this scenario, solutions are sought to replace such methods, with regard to energy storage, mainly focused on renewable energy [89].

Given the existence, in principle and in general, of two main forms of renewable energy generation, the following applies: wind energy (by means of wind turbines) requires a greater investment given the dependence on the installation of large towers with propellers, while solar energy (through photovoltaic panels) is cheaper and has easier installation, since it is installed on a roof or the floor, considering its structure occupying only a few square meters, and requiring only the sunlight [90–92].

To meet an increasing energy demand across the globe, larger batteries with high and long-lasting storage capacities will be widely used in electric and hybrid vehicles. Thus, the use of batteries must also be strongly encouraged in the generation of renewable energy systems for both companies, homes, and even in rural areas which will improve the efficiency of systems and productivity [93].

Energy storage is a solution for isolated systems, without connection to transmission lines and dependent on diesel generators, which are naturally more expensive and polluting. In this context, lithium-ion batteries are evaluated as the main technology that can be used to allow energy to be stored. And in that sense, they promote individual energy self-sufficiency and at the same time cancel the need for supplies by the public network [93–95].

While assessing the difficulty of storing energy generated by renewable sources for those scenarios when there is no sun or wind, the solution is lithium-ion batteries. The technology can allow up to eight hours of energy to be saved and reduce the intermittency in the electrical systems, thus reinforcing the penetration of renewable sources [94, 95].

Regarding developing countries like Brazil, which is one of the countries with the highest incidence of sunlight in the world, greater adoption is expected for the generation of solar energy. The acceptance of solar energy with consumers is growing, reflecting the continuing fall in costs of acquisition and maintenance of photovoltaic systems.

The technologies for the storage of energy in Smart Grids appear as an innovative option, which in addition to reducing costs and environmental impact, increases energy generation and minimizes the incidence of failures. The decentralization of the energy system in smart grids provides sustainable solution for wind and solar energy, and it can be present in any home, i.e., anyone can produce energy and store or sell the surplus [95].

There is a possibility of integrating renewable energy with intelligent energy storage and power generation using diesel or conventional gas, creating a hybrid structure. The technology of Microgrids can be used for local networks as their properties and characteristics are suitable to serve some houses or neighborhoods, allowing consumers to be partially or totally disconnected from a given electricity network of a certain concessionaire. The technology contains photovoltaic solar modules and energy storage, providing autonomous operation capacity using locally generated energy in the event of a power outage from the utility network [94, 95].

Microgrid technology is a reality that corresponds to the possibility of generating local energy and increasing the storage capacity in batteries. In other words, based on the premise that the more renewable energy connected to the local network, the more storage becomes a key part with respect to less dependence on the concessionaire's electrical network, given the better storage capacity. This reflects an increase in efficiency without dependence on the network and an ideal total cost of ownership, considering desired cost savings, reflecting the control and operation management of the microgrid system. These can also mitigate possible problems through remote fault diagnosis, opening opportunities for operational improvements. Microgrids are a great alternative in remote regions that require investment in infrastructure for transmission and distribution lines. This provides an efficient energy setup for local consumers without transmission lines and transformer losses. These systems also have high performance and they are a scalable system. They can be planned, designed and installed quickly even in challenging environments [93, 95].

References

1. Cebulla, Felix, Tobias Naegler, and Markus Pohl. "Electrical energy storage in highly renewable European energy systems: capacity requirements, spatial distribution, and storage dispatch." *Journal of Energy Storage* 14 (2017): 211-223.

2. Kittner, Noah, Felix Lill, and Daniel M. Kammen. "Energy storage deployment and innovation for the clean energy transition." *Nature Energy* 2.9 (2017): 17125.
3. Neto, Pedro Bezerra Leite, Osvaldo R. Saavedra, and Luiz Antonio de Souza Ribeiro. "A dual-battery storage bank configuration for isolated microgrids based on renewable sources." *IEEE Transactions on Sustainable Energy* 9.4 (2018): 1618-1626.
4. Belmonte, N. A. D. I. A., et al. "A comparison of energy storage from renewable sources through batteries and fuel cells: A case study in Turin, Italy." *International Journal of Hydrogen Energy* 41.46 (2016): 21427-21438.
5. Jing, Wenlong, et al. "A comprehensive study of battery-supercapacitor hybrid energy storage system for standalone PV power system in rural electrification." *Applied Energy* 224 (2018): 340-356.
6. Gür, Turgut M. "Review of electrical energy storage technologies, materials and systems: challenges and prospects for large-scale grid storage." *Energy & Environmental Science* 11.10 (2018): 2696-2767.
7. Li, Yang, et al. "Optimal distributed generation planning in active distribution networks considering integration of energy storage." *Applied Energy* 210 (2018): 1073-1081.
8. Rostami, Mohammadali, and Saeed Lotfifard. "Scalable coordinated control of energy storage systems for enhancing power system angle stability." *IEEE Transactions on Sustainable Energy* 9.2 (2017): 763-770.
9. Babatunde, O. M., J. L. Munda, and Y. Hamam. "Power system flexibility: A review." *Energy Reports* 6 (2020): 101-106.
10. Colbertaldo, Paolo, et al. "Impact of hydrogen energy storage on California electric power system: Towards 100% renewable electricity." *International Journal of Hydrogen Energy* 44.19 (2019): 9558-9576.
11. Freeman, James, et al. "A small-scale solar organic Rankine cycle combined heat and power system with integrated thermal energy storage." *Applied Thermal Engineering* 127 (2017): 1543-1554.
12. Yao, Liangzhong, et al. "Challenges and progresses of energy storage technology and its application in power systems." *Journal of Modern Power Systems and Clean Energy* 4.4 (2016): 519-528.
13. Zsiborács, Henrik, et al. "Intermittent renewable energy sources: The role of energy storage in the European power system of 2040." *Electronics* 8.7 (2019): 729.
14. Leonard, Matthew D., Efstathios E. Michaelides, and Dimitrios N. Michaelides. "Energy storage needs for the substitution of fossil fuel power plants with renewables." *Renewable Energy* 145 (2020): 951-962.
15. Liu, Huan, Xiaodong Wang, and Dezhen Wu. "Innovative design of microencapsulated phase change materials for thermal energy storage and versatile applications: a review." *Sustainable Energy & Fuels* 3.5 (2019): 1091-1149.

16. Shen, Xinwei, et al. "Expansion planning of active distribution networks with centralized and distributed energy storage systems." *IEEE Transactions on Sustainable Energy* 8.1 (2016): 126-134.
17. Thapa, Suman, and Rajesh Karki. "Reliability benefit of energy storage in wind integrated power system operation." *IET Generation, Transmission & Distribution* 10.3 (2016): 807-814.
18. Strbac, Goran, et al. "Opportunities for energy storage: assessing whole-system economic benefits of energy storage in future electricity systems." *IEEE Power and Energy Magazine* 15.5 (2017): 32-41.
19. Kyriakopoulos, Grigorios L., and Garyfallos Arabatzis. "Electrical energy storage systems in electricity generation: Energy policies, innovative technologies, and regulatory regimes." *Renewable and Sustainable Energy Reviews* 56 (2016): 1044-1067.
20. Kim, Haein, and Tae Yong Jung. "Independent solar photovoltaic with Energy Storage Systems (ESS) for rural electrification in Myanmar." *Renewable and Sustainable Energy Reviews* 82 (2018): 1187-1194.
21. Liu, Ju, et al. "Stability analysis and energy storage-based solution of wind farm during low voltage ride through." *International Journal of Electrical Power & Energy Systems* 101 (2018): 75-84.
22. Ahmadi, Mikaeel, et al. "Optimal multi-configuration and allocation of SVR, capacitor, centralised wind farm, and energy storage system: a multi-objective approach in a real distribution network." *IET Renewable Power Generation* 13.5 (2019): 762-773.
23. Akbari, Hoda, et al. "Efficient energy storage technologies for photovoltaic systems." *Solar Energy* 192 (2019): 144-168.
24. Basu, M. "Multi-region dynamic economic dispatch of solar–wind–hydro–thermal power system incorporating pumped hydro energy storage." *Engineering Applications of Artificial Intelligence* 86 (2019): 182-196.
25. Han, Li, Rongchang Zhang, and Kai Chen. "A coordinated dispatch method for energy storage power system considering wind power ramp event." *Applied Soft Computing* 84 (2019): 105732.
26. Tan, Darren HS, et al. "From nanoscale interface characterization to sustainable energy storage using all-solid-state batteries." *Nature Nanotechnology* (2020): 1-11.
27. Zelinsky, Michael A., John M. Koch, and Kwo-Hsiung Young. "Performance Comparison of Rechargeable Batteries for Stationary Applications (Ni/MH vs. Ni-Cd and VRLA)." *Batteries* 4.1 (2018): 1.
28. Smith, Kandler, et al. "Life prediction model for grid-connected Li-ion battery energy storage system." 2017 *American Control Conference* (ACC). IEEE, 2017.
29. Yang, Fengchang, et al. "Sodium–sulfur flow battery for low-cost electrical storage." *Advanced Energy Materials* 8.11 (2018): 1701991.

30. Sebastián, Rafael. "Battery energy storage for increasing stability and reliability of an isolated Wind Diesel power system." *IET Renewable Power Generation* 11.2 (2016): 296-303.
31. Rehman, Ali, et al. "Integrated biomethane liquefaction using exergy from the discharging end of a liquid air energy storage system." *Applied Energy* 260 (2020): 114260.
32. Mohammadi, Amin, et al. "Exergy analysis of a Combined Cooling, Heating and Power system integrated with wind turbine and compressed air energy storage system." *Energy Conversion and Management* 131 (2017): 69-78.
33. Ye, Ruijie, et al. "Redox flow batteries for energy storage: a technology review." *Journal of Electrochemical Energy Conversion and Storage* 15.1 (2018).
34. Yan, Yi, et al. "An integrated design for hybrid combined cooling, heating and power system with compressed air energy storage." *Applied Energy* 210 (2018): 1151-1166.
35. Yan, Ruiting, and Qing Wang. "Redox - Targeting - Based Flow Batteries for Large - Scale Energy Storage." *Advanced Materials* 30.47 (2018): 1802406.
36. Ahmadi, Leila, et al. "A cascaded life cycle: reuse of electric vehicle lithium-ion battery packs in energy storage systems." *International Journal of Life Cycle Assessment* 22.1 (2017): 111-124.
37. Adrees, Atia, Hooman Andami, and Jovica V. Milanović. "Comparison of dynamic models of battery energy storage for frequency regulation in power system." *2016 18th Mediterranean Electrotechnical Conference* (MELECON). IEEE, 2016.
38. Nikolić, Zoran, and Dušan Nikolić. "Practical Example Of Solar And Hydro Energy Hybrid System-The Need for a Reversible Power Plant." *Zbornik Međunarodne konferencije o obnovljivim izvorima električne energije–MKOIEE* 6.1 (2018): 165-171.
39. Zhang, Ning, et al. "Reducing curtailment of wind electricity in China by employing electric boilers for heat and pumped hydro for energy storage." *Applied Energy* 184 (2016): 987-994.
40. Arani, AA Khodadoost, et al. "Review of Flywheel Energy Storage Systems structures and applications in power systems and microgrids." *Renewable and Sustainable Energy Reviews* 69 (2017): 9-18.
41. Faraji, Faramarz, Abbas Majazi, and Kamal Al-Haddad. "A comprehensive review of flywheel energy storage system technology." *Renewable and Sustainable Energy Reviews* 67 (2017): 477-490.
42. Morstyn, Thomas, Martin Chilcott, and Malcolm D. McCulloch. "Gravity energy storage with suspended weights for abandoned mine shafts." *Applied Energy* 239 (2019): 201- 206.
43. Hou, Hui, et al. "Optimal capacity configuration of the wind-photovoltaic-storage hybrid power system based on gravity energy storage system." *Applied Energy* 271 (2020): 115052.

44. Nasri, Sihem, Ben Slama Sami, and Adnane Cherif. "Power management strategy for hybrid autonomous power system using hydrogen storage." *International Journal of Hydrogen Energy* 41.2 (2016): 857-865.
45. Benato, Alberto. "Performance and cost evaluation of an innovative Pumped Thermal Electricity Storage power system." *Energy* 138 (2017): 419-436.
46. Moreno, Rodrigo, et al. "Facilitating the integration of renewables in Latin America: the role of hydropower generation and other energy storage technologies." *IEEE Power and Energy Magazine* 15.5 (2017): 68-80.
47. Prasad, J. Sunku, et al. "A critical review of high-temperature reversible thermochemical energy storage systems." *Applied Energy* 254 (2019): 113733.
48. Arenas, L. F., C. Ponce de León, and F. C. Walsh. "Engineering aspects of the design, construction and performance of modular redox flow batteries for energy storage." *Journal of Energy Storage* 11 (2017): 119-153.
49. Xing, Xuetao, et al. "Time-Varying Model Predictive Control of a Reversible-SOC Energy-Storage Plant based on the Linear Parameter-Varying Method." *IEEE Transactions on Sustainable Energy* (2019).
50. Wen, Shuli, et al. "Optimal sizing of hybrid energy storage sub-systems in PV/diesel ship power system using frequency analysis." *Energy* 140 (2017): 198-208.
51. Li, Jianwei, et al. "Design/test of a hybrid energy storage system for primary frequency control using a dynamic droop method in an isolated microgrid power system." *Applied Energy* 201 (2017): 257-269.
52. Kåberger, Tomas. "Progress of renewable electricity replacing fossil fuels." *Global Energy Interconnection* 1.1 (2018): 48-52.
53. Lazkano, Itziar, Linda Nøstbakken, and Martino Pelli. "From fossil fuels to renewables: The role of electricity storage." *European Economic Review* 99 (2017): 113-129.
54. Shakouri, Bahram, and Soheila Khoshnevis Yazdi. "Causality between renewable energy, energy consumption, and economic growth." *Energy Sources, Part B: Economics, Planning, and Policy* 12.9 (2017): 838-845.
55. Faria, A. F., et al. "Energy efficiency and renewable energy: Energy, economics and environment gains." *2017 IEEE URUCON*. IEEE, 2017.
56. Kaabeche, A., and Y. Bakelli. "Renewable hybrid system size optimization considering various electrochemical energy storage technologies." *Energy Conversion and Management* 193 (2019): 162-175.
57. Yang, Bo, et al. "A Durable, Inexpensive and Scalable Redox Flow Battery Based on Iron Sulfate and Anthraquinone Disulfonic Acid." *Journal of the Electrochemical Society* 167.6 (2020): 060520.
58. Maheshwari, Arpit, et al. "Optimizing the operation of energy storage using a non-linear lithium-ion battery degradation model." *Applied Energy* 261 (2020): 114360.
59. Zhao, Zequan, et al. "Challenges in zinc electrodes for alkaline zinc-air batteries: obstacles to commercialization." *ACS Energy Letters* 4.9 (2019): 2259-2270.

60. Yao, Erren, et al. "Thermo-economic optimization of a combined cooling, heating and power system based on small-scale compressed air energy storage." *Energy Conversion and Management* 118 (2016): 377-386.
61. Krawczyk, Piotr, et al. "Comparative thermodynamic analysis of compressed air and liquid air energy storage systems." *Energy* 142 (2018): 46-54.
62. Nikiforidis, Georgios, M. C. M. Van De Sanden, and Michail N. Tsampas. "High and intermediate temperature sodium-sulfur batteries for energy storage: development, challenges and perspectives." *RSC Advances* 9.10 (2019): 5649-5673.
63. Mehrpooya, Mehdi, and Pouria Pakzad. "Introducing a hybrid mechanical–Chemical energy storage system: Process development and energy/exergy analysis." *Energy Conversion and Management* 211 (2020): 112784.
64. Aktas, Ahmet, et al. "Dynamic energy management for photovoltaic power system including hybrid energy storage in smart grid applications." *Energy* 162 (2018): 72-82.
65. Nikolaidis, Pavlos, and Andreas Poullikkas. "Cost metrics of electrical energy storage technologies in potential power system operations." *Sustainable Energy Technologies and Assessments* 25 (2018): 43-59.
66. Rocabert, Joan, et al. "Control of energy storage system integrating electrochemical batteries and supercapacitors for grid-connected applications." *IEEE Transactions on Industry Applications* 55.2 (2018): 1853-1862.
67. Breeze, Paul. *Power system energy storage technologies.* Academic Press, 2018.
68. Al-Ghussain, Loiy, et al. "Sizing renewable energy systems with energy storage systems in microgrids for maximum cost-efficient utilization of renewable energy resources." *Sustainable Cities and Society* 55 (2020): 102059.
69. Luo, Liang, et al. "Optimal scheduling of a renewable-based microgrid considering photovoltaic system and battery energy storage under uncertainty." *Journal of Energy Storage* 28 (2020): 101306.
70. Liu, Ye, et al. "Optimal sizing of a wind-energy storage system considering battery life." *Renewable Energy* 147 (2020): 2470-2483.
71. Yan, Ning, et al. "Economic dispatch application of power system with energy storage systems." *IEEE Transactions on Applied Superconductivity* 26.7 (2016): 1-5.
72. Nguyen, Tu A., and Raymond H. Byrne. "Maximizing the cost-savings for time-of-use and net-metering customers using behind-the-meter energy storage systems." *2017 North American Power Symposium* (NAPS). IEEE, 2017.
73. Abdin, Giulio Cerino, and Michel Noussan. "Electricity storage compared to net metering in residential PV applications." *Journal of Cleaner Production* 176 (2018): 175-186.

74. Moazzami, Majid, et al. "Optimal Economic Operation of Microgrids Integrating Wind Farms and Advanced Rail Energy Storage System." *International Journal of Renewable Energy Research* (IJRER) 8.2 (2018): 1155-1164.
75. Walawalkar, Rahul. "Advances in long-duration storage technologies." *IASH Journal- International Association for Small Hydro* 9.1 (2019): 37-38.
76. Aghahosseini, Arman, and Christian Breyer. "Assessment of geological resource potential for compressed air energy storage in global electricity supply." *Energy Conversion and Management* 169 (2018): 161-173.
77. Hubble, Andrew Harrison, and Taha Selim Ustun. "Scaling renewable energy-based microgrids in underserved communities: Latin America, South Asia, and Sub-Saharan Africa." *2016 IEEE PES PowerAfrica*. IEEE, 2016.
78. Li, Yunming, et al. "Recent advances of electrode materials for low-cost sodium-ion batteries towards practical application for grid energy storage." *Energy Storage Materials* 7 (2017): 130-151.
79. De Rosa, Mattia, Mark Carragher, and Donal P. Finn. "Flexibility assessment of a combined heat-power system (CHP) with energy storage under real-time energy price market framework." *Thermal Science and Engineering Progress* 8 (2018): 426-438.
80. Li, Fan, et al. "A hybrid optimization-based scheduling strategy for combined cooling, heating, and power system with thermal energy storage." *Energy* 188 (2019): 115948.
81. Rao, Zhonghao, et al. "Experimental study on a novel form-stable phase change materials based on diatomite for solar energy storage." *Solar Energy Materials and Solar Cells* 182 (2018): 52-60.
82. Adibhatla, Sairam, and S. C. Kaushik. "Energy, exergy, economic and environmental (4E) analyses of a conceptual solar aided coal-fired 500 MWe thermal power plant with thermal energy storage option." *Sustainable Energy Technologies and Assessments* 21 (2017): 89-99.
83. Ma, Hengrui, et al. "Optimal scheduling of an regional integrated energy system with energy storage systems for service regulation." *Energies* 11.1 (2018): 195.
84. Kumar, Sanjiv, Neeta Kumar, and Saxena Vivekadhish. "Millennium development goals (MDGs) to sustainable development goals (SDGs): Addressing unfinished agenda and strengthening sustainable development and partnership." *Indian Journal of Community Medicine: official publication of Indian Association of Preventive & Social Medicine* 41.1 (2016): 1.
85. Singh, Yaduvir, and Nitai Pal. "Obstacles and comparative analysis in the advancement of photovoltaic power stations in India." *Sustainable Computing: Informatics and Systems* 25 (2020): 100372.
86. Murty, V. V. S. N., and Ashwani Kumar. "Multi-objective energy management in microgrids with hybrid energy sources and battery energy

storage systems." *Protection and Control of Modern Power Systems* 5.1 (2020): 1-20.
87. Hou, Jun, et al. "Control Strategy for Battery/Flywheel Hybrid Energy Storage in Electric Shipboard Microgrids." *IEEE Transactions on Industrial Informatics* (2020).
88. Li, Yaowang, et al. "Dynamic modelling and techno-economic analysis of adiabatic compressed air energy storage for emergency back-up power in supporting microgrid." *Applied Energy* 261 (2020): 114448.
89. Simla, Tomasz, and Wojciech Stanek. "Reducing the impact of wind farms on the electric power system by the use of energy storage." *Renewable Energy* 145 (2020): 772-782.
90. Zhao, Pan, et al. "Technical feasibility assessment of a standalone photovoltaic/wind/adiabatic compressed air energy storage based hybrid energy supply system for rural mobile base station." *Energy Conversion and Management* 206 (2020): 112486.
91. Diyoke, Chidiebere, and Chunfei Wu. "Thermodynamic analysis of hybrid adiabatic compressed air energy storage system and biomass gasification storage (A-CAES+ BMGS) power system." *Fuel* 271 (2020): 117572.
92. Zhang, Zehui, Cong Guan, and Zhiyong Liu. "Real-Time Optimization Energy Management Strategy for Fuel Cell Hybrid Ships Considering Power Sources Degradation." *IEEE Access* 8 (2020): 87046-87059.
93. Kamath, Dipti, et al. "Economic and Environmental Feasibility of Second-Life Lithium- Ion Batteries as Fast-Charging Energy Storage." *Environmental Science & Technology* 54.11 (2020): 6878-6887.
94. Diahovchenko, Illia, et al. "Progress and Challenges in Smart Grids: Distributed Generation, Smart Metering, Energy Storage and Smart Loads." *Iranian Journal of Science and Technology, Transactions of Electrical Engineering* (2020): 1-15.
95. Aslam, Sheraz, Adia Khalid, and Nadeem Javaid. "Towards efficient energy management in smart grids considering microgrids with day-ahead energy forecasting." *Electric Power Systems Research* 182 (2020): 106232.

3
Virtual Inertia Provision Through Energy Storage Technologies

Shreya Mahajan* and Yajvender Pal Verma

Department of EEE, UIET, Panjab University, Chandigarh, India

Abstract

With the rise in utilisation level of renewables, the effect of small inertia changes the dynamics and stability of the system network. A possible solution with regard to stabilization of such a grid is to provide virtual inertia in the system. This chapter discusses the concept of virtual inertia to improve the frequency control in a grid with a large share of renewable energy sources. The impact of low system inertia on power system voltage and operation and control due to a large share of renewables has been highlighted in detail. Virtual inertia emulation methods for wind and solar PV systems without and with energy storage systems have been elaborated in this chapter. The utilisation of various energy storage technologies for emulating virtual inertia has been discussed.

Keywords: Virtual inertia, photovoltaic, energy storage technologies, frequency regulation, battery energy storage

3.1 Introduction

Worldwide power demand is increasing rapidly; it is almost doubling in most developing countries in every ten years [1]. Therefore, the modern power system is being expanded and interconnected to meet the rising electric power demand. The modern power system mainly comprises

---

*Corresponding author: 123shreya.m@gmail.com

Sandeep Dhundhara and Yajvender Pal Verma (eds.) Energy Storage for Modern Power System Operations, (59–78) © 2021 Scrivener Publishing LLC

conventional sources as well as Renewable Energy Sources (RESs). The major proportion of power demand is met by conventional energy sources like thermal, hydro, nuclear, gas turbines, etc. [2–4].

However, the gradual depletion of fossil fuels, electricity market liberalization and demand for cheaper, reliable and eco-friendly power have encouraged the widespread utilization of the RESs. Use of RESs has been increasing rapidly with the degradation of fossil fuels and due to the negative impacts of fossil fuels on the environment. As renewable energy sources are unpredictable, it is a difficult task to incorporate RES into the power system and this results in many technical and non-technical challenges [5]. The rise in integration of Photovoltaic (PV) and wind power in a power system is increasing the concerns due to the negative effect of the power fluctuations produced in these systems [6]. There are numerous challenges and problems linked with the incorporation of various RESs such as solar PV and wind power, which are described below [7–9]:

1. Power quality issues
    a. Harmonics
    b. Voltage instability
    c. Frequency instability
2. Power variations
3. Storage challenges
4. Sizing and location of RESs.

Operating characteristics of RESs are different than the conventional generation of electricity and as such they do not provide any system inertia [10]. This inertia is one of the most crucial functioning parameters which decide the frequency stability of the system. In the event of any fault which results in frequency stability, alternators will inject or absorb kinetic energy into or from the grid to eliminate the frequency changes [11]. If the system inertia is lower, the system is vulnerable to more frequency deviations which lead to network stability problem. Wind turbines such as Doubly Fed Induction Generators (DFIG) provide no inertia in the event of frequency fluctuations even though enough kinetic energy is stored in its blades and the generator [12, 13]. Solar PV panels do not have any inertia in the system. Therefore, less inertia leads to the lower damping in the system and more deviations in grid frequency. The speedy growth of RESs is shifting the power system towards an inverter-dominated system as depicted in Figure 3.1.

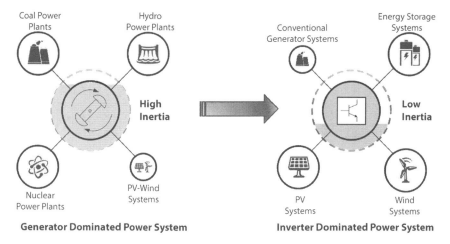

Figure 3.1 Traditional v/s future RES-based inverter dominated power system [14].

3.2 Virtual Inertia-Based Frequency Control

With the increased utilisation of RESs, the effect of low inertia and smaller damping is raising numerous issues in power network performance and grid efficiency. Excessive voltage drop in the system [15], power fluctuations as a result of intermittent behaviour of RESs and poor frequency response are a few negative issues of the related problem. A possible solution of the above-mentioned problems is to provide additional virtual inertia. A virtual inertia may be provided for DGs/RESs by utilising short-term energy storage in conjunction with an inverter and an appropriately designed control system. This mechanism is called Virtual Synchronous Generator (VSG) or Virtual Synchronous Machine (VSM) [16]. The VSM is a combination of Voltage Source Converters (VSC) and Synchronous Machine (SM) properties. The VSM control loop of a power inverter is mainly a control system which is added to the inverter controller so that it behaves like an SM. The voltage converter has no inertia or damping behaviour, unlike SMs.

3.2.1 Concept of Virtual Inertia

The main aim of the VSM is to provide dynamic characteristics of a synchronous machine for the power electronics–based DGs and RESs so as to mimic the behaviour of an SM in maintaining grid stability. The concept of the VSM may be applied either to a single DG or multiple DGs in the

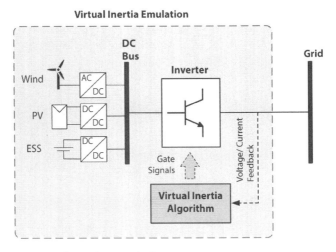

Figure 3.2 Concept of virtual inertia [19].

system [17]. The VSM comprises an energy storage system, power inverter and a control loop as illustrated in Figure 3.2 [18]. The VSM is situated between a DC bus and the power grid. The virtual inertia is emulated by controlling the real power through the power inverter in inverse proportion of the rotor speed.

3.2.2 Virtual Inertia Emulation

For improving the frequency response in the system network with a huge share of RESs, the following control techniques have been used in this area of research:

- Energy Storage System (ESS) such as batteries or capacitors is linked with a PV unit or wind turbine [20]. An appropriate control mechanism is required between the ESS and the wind turbine/PV unit for optimisation of power produced by the renewables for maintaining frequency regulation.
- By deloading of wind farm or PV unit, the units work at a sub-optimum operational working point and a power reserve is produced which is used for frequency regulation [21].
- Demand Side Management (DSM) is also incorporated at the load end. In [22], a control technique using DSM has been proposed for frequency stability.

3.3 Impact of Low System Inertia on Power System Voltage and Operation & Control Due to Large Share of Renewables

A large share of RESs in the distribution and transmission system has both positive and negative effects on the normal functioning of the system. Improvement in voltage profile, power loss minimisation and other ancillary services support are some positive impacts. However, negative effects involve protection system failure, poor power quality, frequency instability, etc. [23, 24]. While there are numerous benefits of deploying DGs in the system, research has indicated that there may be an increase in system losses due to improper location and size of DGs. By including optimization techniques, it is possible to rectify the problems of power losses and voltage regulation [25]. The literature surveys related to the problem of optimal deployment of DG have been highlighted in [26–28] and various methodologies for optimal deployment of DG can be classified as (a) classical (b) analytical, and (c) meta-heuristic and heuristic. Some of these methods are: Artificial Bee Colony (ABC) [29], Modified Honey Bee Mating [30], Krill Herd Algorithm (KHA) [31], Improved PSO [32]. Table 3.1 summarizes the impact of low inertia in a power system network. An optimization strategy related to Grey Wolf Optimizer (GWO) for deployment of multiple DG units in the distribution system has been proposed in [33]. In [34], a new optimization technique is proposed that is based on flower pollination algorithm to determine the optimum DG size and location. Maintenance of system frequency within an acceptable range is the foremost condition for the stable operation of a system network. A framework to analyse the impact of different penetration levels of RESs and its effect on lowering of inertia and deviations in grid frequency has been presented in [35, 36]. The developed methodology is demonstrated using three operating conditions of the network. An extensive literature review on the Load Frequency Control (LFC) problem in a power system has been highlighted in [37].

In [38], a frequency and voltage control approach for a stand-alone microgrid with a large share of RESs has been suggested and the impact of low inertia on frequency and voltage deviations has been studied. A Battery Energy Storage System (BESS) has been incorporated for generating the nominal system frequency and for maintaining frequency stability. Jaewan Suh *et al.* in [39] proposed a frequency control strategy with a high share of renewables for attaining flexibility in the operation of power grid. Ebrahim Rokrok *et al.* in [40] reviewed numerous primary frequency control methods which have been proposed to control the voltage and

Table 3.1 Summary of impact of low inertia on power system.

References	Objectives	Findings of the work done
[25]	Study of issues with penetration of DR into power system.	Effect of DG on system operation such as voltage fluctuations, high power losses, poor power quality, failure of protection, frequency reliability and regulation issues.
[26–28]	Review on optimum deployment and selection of appropriate size for single and multiple DG units.	Deployment of RES studies has indicated that there may be an increase in system losses due to improper location and size of DGs.
[29, 31–34]	Artificial intelligence and nature inspired DG integration techniques for improving the system profile.	• Artificial Bee Colony (ABC) [29] • Krill Herd Algorithm (KHA) [31] • PSO [32] • Grey Wolf Optimizer (GWO) for deployment of multiple DG units [33] • Flower pollination algorithm for optimal location and size of DG and RES [34]
[35, 36]	Analysis of the effect of various penetration levels of RES and its effect on lowering of inertia and deviations in grid frequency.	• Reduction in inertia limits of the system for frequency stability • Highlights the effect of primary frequency response and impact of low inertia on frequency stability.

(Continued)

Table 3.1 Summary of impact of low inertia on power system. (*Continued*)

References	Objectives	Findings of the work done
[38]	• Frequency and voltage control approach for microgrid with large share of RESs. • Battery Energy Storage System (BESS) has been incorporated for producing optimum frequency and for maintaining frequency stability.	Proposed (Q/P) droop control alleviates the voltage fluctuations produced by its own output power fluctuations.
[40, 41]	Literature review on load frequency control (LFC) in distribution system network.	Investigations on LFC problems and issued with usage of storage devices like BESS, FACTS devices in wind and PV systems.
[42]	• Voltage control loop controlled by automatic voltage regulator of SM. • Optimisation using Cuckoo search algorithm.	Voltage and frequency deviation minimisation.

frequency of microgrids. The design of a supervisory control strategy for a grid-connected wind farm is presented in [41]. In [42], Wenlei Bai *et al.* have suggested a voltage control loop controlled by automatic voltage regulator of SM. This voltage control loop is combined with frequency control loop to attain optimum response for frequency and voltage deviations. In this research paper, Cuckoo Search Algorithm has been used for optimisation of gain to minimize frequency and voltage variations.

3.4 Control Methods for Inertia Emulation in RES-Based Power Systems

There are numerous inertia and frequency control techniques for RESs with and without using ESS to counter the negative effects of low inertia with a larger share of renewables. These techniques as shown in Figure 3.3

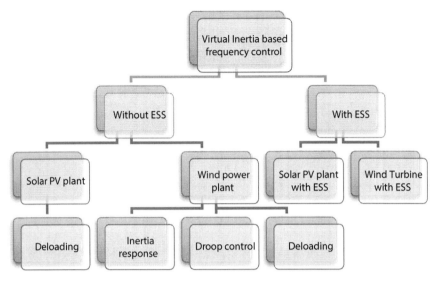

Figure 3.3 Virtual inertia emulation methods for wind and solar PV systems.

contribute to the frequency stability with high penetration of RESs such as PV and wind power.

3.4.1 Control Methods Without ESS for Frequency Control

First, the literature review on the control techniques designed for RESs without energy storage system has been presented in Table 3.2. There are mainly two techniques to support frequency regulation using a wind turbine: inertia response and power reserve control. In the inertia control method, the wind turbine is made to release the kinetic energy stored in the rotating blades in the range of 10-15 seconds to control the frequency variations. In the reserve control technique, a pitch angle controller or speed controller is used to increase the power reserve during faulty events. These methods are elaborately presented in literature [43, 44].

The droop control scheme, proposed by S. Mishra *et al.* in [45] regulates the active power output obtained from wind system which is proportional to frequency change and helps in improving frequency regulation and stability of the system. Dooyeon Kim *et al.* [46] suggested a virtual inertial control for a wind power–based system based on the frequency deviation.

Presently, there has been an increase in the share of PV in distribution systems, as a result of which reserve power from the traditional sources of energy is not enough to maintain frequency stability. To participate in frequency regulation, different modifications have been done in designing of controllers which involves a larger share of PV units in the system. The deloading technique for solar PV without ESS has been discussed in [47, 48] as shown in Figure 3.4. These papers highlight comprehensive control strategy that enables the solar PV system to regulate frequency. L.D. Watson *et al.* in [49] discussed a control scheme proposed for a PV panel to regulate the frequency deviations of an islanded microgrid.

Researchers are also working on the use of a DC link capacitor of inverter to extract the kinetic energy and stored energy for the purpose of overcoming the problem of low inertia in RES. In [50, 51], the virtual inertia control techniques have been introduced that mimic the working inertia of synchronous generator to improve frequency regulation.

The authors in [52, 53] suggest the idea of virtual inertia which can be emulated by the energy stored in the dc-link capacitors of inverters. The dc-link capacitors are accumulated as a large equivalent capacitor acting like energy buffers for frequency control and regulation. In literature [54] and [55], an unconventional method is suggested to increase the inertia of a PV system through inertia emulation as per the topology as shown in Figure 3.5. The inertia emulation is done by controlling the charging and discharging of the DC link capacitor and with the adjustment of the PV generation. Table 3.2 summarises the virtual inertia emulation techniques without ESS.

3.4.2 Control Methods with ESS for Frequency Control

In the previous section, numerous control techniques have been discussed to provide RES with the ability to emulate inertia virtually for frequency regulation. Conversely, there have been some limitations of using these techniques in terms of system reliability due to the intermittent nature of renewables in the power system. In order to reduce those drawbacks, ESSs must be added to wind turbines and solar PVs to improve and maintain frequency regulation. This literature review presents the detailed control methods for virtual inertia emulation using ESS as shown in Figure 3.6. This section deals with nonconventional technologies such as Battery Energy Storage System (BESS), pumped hydroelectric energy storage, compressed air energy storage, flywheels energy storage system (FESS), supercapacitors and ultra-capacitors and hybrid energy storage systems (HESS) that can provide better and fast response for maintaining power balance. Table 3.3 summarizes the energy storage technologies used for virtual inertia emulation.

Table 3.2 Summary of virtual inertia emulation techniques without ESS.

Ref.	Types of source	Methodology	Objectives of the work done	Findings
[44]	Variable speed wind turbine	Inertia response	variable-speed wind turbines emulating inertia for frequency control.	Power reserve is increased to emulate inertia and to maintain frequency control.
[50]	DFIG	Droop control	To emulate inertia in DFIG for improving system stability	Benefits and drawbacks of using DFIG to emulate virtual inertia.
[51]	DFIG	Deloading technique	Comparison of the permanent magnet synchronous generator and DFIG to emulate inertia in the system.	Reduction of ROCOF and frequency deviations.
[52]	DFIG	DC link capacitor	distributed virtual inertia which can be emulated by the energy in the dc-link capacitors of inverters.	The dc-link capacitors are accumulated as a large equivalent capacitor provides energy buffer for frequency control and regulation.
[45, 47–49]	Solar PV plant	Deloading technique	Frequency control by deloaded PV.	Reduction in frequency deviations.
[54, 55]	Solar PV plant	DC link capacitor	To enhance the inertia of PV systems by emulation of inertia.	controlling the charging/discharging of the DC link capacitor and with the adjustment of the PV generation.

VIRTUAL INERTIA PROVISION USING ENERGY STORAGE TECHNOLOGIES 69

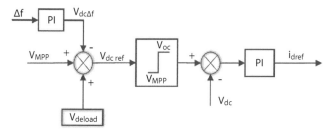

Figure 3.4 Deloading controller for Solar PV plant [56].

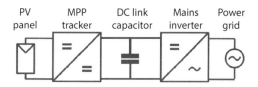

Figure 3.5 Topology for DC link capacitor in PV system [54].

3.4.2.1 Battery Energy Storage Systems (BESS)

Batteries are energy storage devices which are used for inertia emulation to offer inertia in power systems with a large share of RES. A coordinated control mechanism with BESS improves the frequency stability of the system network. A lithium-ion battery is commonly used in a low-inertia power system which helps in improving frequency stability. The frequency and voltage regulation control methods in a PV system with lithium-ion batteries is presented in [56]. In [57], Manohar Chamana *et al.* proposed a control methodology where a PV source is connected with BESS at the DC link bus. This provides an advantage in that it can emulate behaviour of synchronous

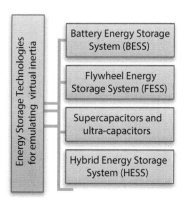

Figure 3.6 Energy storage technologies for emulating virtual inertia.

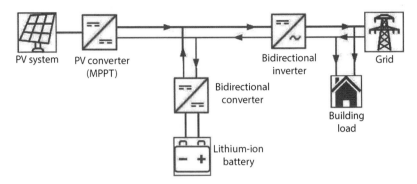

Figure 3.7 BESS-based frequency regulation in PV system [57].

generator as well as contribute in providing reserve power. The incorporation of BESS allows PV units with the ability to control frequency and voltage regulation when it is connected to the grid as shown in Figure. 3.7. Li Xin and Wang Ning in [58] suggested a concept of virtual synchronous machine with ESS and its control mechanism for frequency control and regulation. In this technique, the power inverter has the ability of power control and frequency regulation that provides system stability when connected to the power grid.

3.4.2.2 *Super Capacitors and Ultra-Capacitors*

Super capacitors and ultra-capacitors are energy storage units which are used to store electric charge and voltage as well as discharge a large quantity of power in less time. Therefore, the faster response time of super capacitors and ultra-capacitors is utilised in inertial response. The literature in [59] proposed a combined VIE and a super capacitor–based energy storage system to improve the frequency stability of the microgrids and for reducing short-term power fluctuations in the grid. The proposed control strategy helped in overcoming the slow response of batteries by moving the power fluctuations to the super capacitor system and improved the frequency regulation by virtually emulating the inertia of synchronous machine. A new control mechanism of super capacitors for enhancing the frequency stability of the PV based microgrid is suggested in [60].

3.4.2.3 *Flywheel Energy Storage System (FESS)*

The flywheel system is a mechanical storage system which stores energy in the form of kinetic energy. The rotor of flywheel is accelerated or decelerated by a motor for absorbing the power and these are connected to the grid to provide inertia. Flywheels are used for providing ancillary services such as kinetic inertia and for maintaining frequency regulation [61]. Flywheels are connected to the grid through a power electronic device. The advantages of a flywheel are fast response, good efficiency and longer life. To solve the lack of inertia issue, [62] proposed the method of using FESS for delivering virtual inertia and for maintaining frequency support. In comparison with batteries, flywheels have a longer lifetime and better frequency regulation behaviour.

3.4.2.4 Hybrid Energy Storage System (HESS)

Although the use of a lithium-battery in a low-inertia power system with a large share of renewables improves frequency stability but there is fast discharge of power from the battery that increases the stress on it. In order to decrease the pressure on the battery and to increase the lifespan of a lithium battery, the combination of an ultra-capacitor and lithium-ion battery are incorporated to provide virtual inertia for maintaining the stability of the power system with high penetration of RESs [63]. In [64], it was presented that the integration of combination of a battery and ultra-capacitor easily managed power fluctuations as shown by the topology in Figure 3.8. The ultra-capacitor manages rapid fluctuations of power and the battery helps in improving frequency regulation.

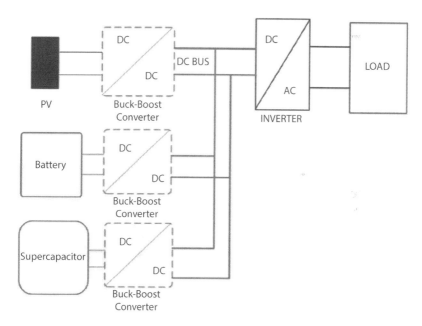

Figure 3.8 HESS for frequency regulation in PV system.

Table 3.3 Summary of energy storage technologies used for virtual inertia emulation.

Ref.	Types of storage	Objectives of the work done	Advantages	Disadvantages
[56–58]	BESS	• To enhance frequency stability of power system.	• Fast response time • Manage sudden power imbalance	• Charging and discharge of battery • Rising cost of energy storage system
[61, 62]	FESS	• To improvise the inertia of system and to minimise the frequency fluctuations.	• Ramping capability • Roundtrip efficiency • Lifetime	• Energy storage capability • Amount of energy it can store
[59, 60]	Super capacitors and ultra-capacitors	• To improve the frequency regulation of the micro grids and for reducing short-term power fluctuations in the grid.	• To improve the frequency stability of the micro grids and for reducing short-term power deviations in the grid. • Fast response time	• Not well-suited for long-term energy storage • Gradual voltage loss
[63, 64]	HESS	• To enhance frequency stability of distribution system network and micro grid.	• Balances sudden power fluctuations.	• Interaction and co-ordination must be controlled and it can become complicated. • High installation cost

3.5 Challenges

Generally, the batteries and capacitors are used as an energy storage system with photovoltaic-based VSGs. However, batteries and conventional capacitor are allowed to release power; therefore, [50–59] proposed a super capacitor–based energy storage system which significantly minimises the release time of stored energy. Although this solution is quite better for energy storage it is not economical due to the high cost. Therefore, it is needed to develop a new and economical type of energy storage system which owns the properties of conventional batteries and ultra-capacitors.

It is essential to develop reliable and effective modelling techniques for VSG by which a large number of VSGs can effectively integrate into the power system without changing crucial power system parameters. Therefore, a detailed, reliable and accurate dynamic model of VSG and power system is required to develop and analyse deeply. The parameter estimation for VSG and power system should be precise and accurate by some specialised and developed algorithms.

The increasing number of DGs into power system enhances the stability issue. Although VSG supplies the inertia to the power system and improves the power system coupling, the overall stability of the power system drops. In the currently available literature, the primary focus of VSG stability problem solving is on individual parts of the system; whereas the stability of the overall system is not researched in depth. Therefore, for future research, it is necessary to develop the control which possesses the features to improve the stability of individual parts as well as the overall system.

References

1. M. Sandhu and T. Thakur, "Issues, Challenges, Causes, Impacts and Utilization of Renewable Energy Sources - Grid Integration," *J. Eng. Res. Appl.*, vol. 4, no. 3, pp. 636–643, g2014.y
2. G. M. Shafiullah, M. T. Arif, and A. M. T. Oo, "Mitigation strategies to minimize potential technical challenges of renewable energy integration," *Sustain. Energy Technol. Assessments*, vol. 25, no. September 2017, pp. 24–42, 2018.
3. A. S. Anees, "Grid integration of renewable energy sources: Challenges, issues and possible solutions," *2012 IEEE 5th India Int. Conf. Power Electron.*, pp. 1–6, 2012.
4. B. Kroposki *et al.*, "Achieving a 100% Renewable Grid: Operating Electric Power Systems with Extremely High Levels of Variable Renewable Energy," *IEEE Power Energy Mag.*, vol. 15, no. 2, pp. 61–73, 2017.

5. A. S. N. Huda and R. Živanović, "Large-scale integration of distributed generation into distribution networks: Study objectives, review of models and computational tools," *Renew. Sustain. Energy Rev.*, vol. 76, no. February, pp. 974–988, 2017.
6. S. U. Karki, S. B. Halbhavi, and S. G. Kulkarni, "Study on Challenges in Integrating Renewable Technologies," *Int. J. Adv. Res. Electr. Electron. Instrum. Eng.*, vol. 3, no. 8, pp. 10972–10977.
7. F. Alsokhiry and K. L. Lo, "Distributed generation based on renewable energy providing frequency response ancillary services," in *4th International Conference on Power Engineering, Energy and Electrical Drives*, 2013, no. May, pp. 1200–1205.
8. P. Kotsampopoulos, N. Hatziargyriou, B. Bletterie, and G. Lauss, "Review, analysis and recommendations on recent guidelines for the provision of ancillary services by Distributed Generation," in *Proceedings - 2013 IEEE International Workshop on Intelligent Energy Systems, IWIES 2013*, 2013, pp. 185–190.
9. A. Banshwar, N. K. Sharma, Y. R. Sood, and R. Shrivastava, "Renewable energy sources as a new participant in ancillary service markets," *Energy Strateg. Rev.*, vol. 18, pp. 106–120, 2017.
10. H. Bevrani, A. Ghosh, and G. Ledwich, "Renewable energy sources and frequency regulation: survey and new perspectives," *IET Renew. Power Gener.*, vol. 4, no. 5, p. 438, 2010.
11. P. Tielens and D. van Hertem, "Grid Inertia and Frequency Control in Power Systems with High Penetration of Renewables," *Status Publ.*, vol. 0, no. 2, pp. 1–6, 2012.
12. P. Tielens and D. Van Hertem, "The relevance of inertia in power systems," *Renew. Sustain. Energy Rev.*, vol. 55, pp. 999–1009, 2016.
13. A. Ulbig, T. S. Borsche, and G. Andersson, "Impact of low rotational inertia on power system stability and operation," *IFAC Proc. Vol.*, vol. 19, pp. 7290–7297, 2014.
14. U. Tamrakar, D. Shrestha, M. Maharjan, B. Bhattarai, T. Hansen, and R. Tonkoski, "Virtual Inertia: Current Trends and Future Directions," *Appl. Sci.*, vol. 7, no. 7, p. 654, 2017.
15. K. Koyanagi, "A smart photovoltaic generation system integrated with lithium-ion capacitor storage," in *46th International Universities' Power Engineering Conference*, 2011.
16. H. Bevrani, T. Ise, and Y. Miura, "Virtual synchronous generators: A survey and new perspectives," *Int. J. Electr. Power Energy Syst.*, vol. 54, pp. 244–254, 2014.
17. K. Visscher and S. W. H. De Haan, "Virtual synchronous machines (VSG'S) for frequency stabilisation in future grids with a significant share of decentralized generation," in *CIRED Seminar 2008: SmartGrids for Distribution*, 2008, no. 118, pp. 82–82.

18. Y. Hirase, K. Abe, K. Sugimoto, and Y. Shindo, "A grid-connected inverter with virtual synchronous generator model of algebraic type," *Electr. Eng. Japan*, vol. 184, no. 4, pp. 10–21, Sep. 2013.
19. Y. Chen, R. Hesse, D. Turschner, and H.-P. Beck, "Improving the grid power quality using virtual synchronous machines," in *2011 International Conference on Power Engineering, Energy and Electrical Drives*, 2011, no. May, pp. 1–6.
20. M. Nayeripour, M. Hoseintabar, and T. Niknam, "Frequency deviation control by coordination control of FC and double-layer capacitor in an autonomous hybrid renewable energy power generation system," *Renew. Energy*, vol. 36, no. 6, pp. 1741–1746, Jun. 2011.
21. X. Yingcheng and T. Nengling, "Review of contribution to frequency control through variable speed wind turbine," *Renew. Energy*, vol. 36, no. 6, pp. 1671–1677, Jun. 2011.
22. I. Serban, C. Marinescu, and C. P. Ion, "A voltage-independent active load for frequency control in microgrids with renewable energy sources," *2011 10th Int. Conf. Environ. Electr. Eng. EEEIC.EU 2011 - Conf. Proc.*, pp. 11–14, 2011.
23. T. Ackermann, G. Andersson, and L. Söder, "Distributed generation: a definition," *Electr. Power Syst. Res.*, vol. 57, no. 3, pp. 195–204, Apr. 2001.
24. M. M. Haque and P. Wolfs, "A review of high PV penetrations in LV distribution networks: Present status, impacts and mitigation measures," *Renew. Sustain. Energy Rev.*, vol. 62, pp. 1195–1208, Sep. 2016.
25. R. A. Walling, R. Saint, R. C. Dugan, J. Burke, and L. A. Kojovic, "Summary of Distributed Resources Impact on Power Delivery Systems," *IEEE Trans. Power Deliv.*, vol. 23, no. 3, pp. 1636–1644, Jul. 2008.
26. Z. Abdmouleh, A. Gastli, L. Ben-Brahim, M. Haouari, and N. A. Al-Emadi, "Review of optimization techniques applied for the integration of distributed generation from renewable energy sources," *Renew. Energy*, vol. 113, pp. 266–280, Dec. 2017.
27. M. Pesaran H.A, P. D. Huy, and V. K. Ramachandaramurthy, "A review of the optimal allocation of distributed generation: Objectives, constraints, methods, and algorithms," *Renew. Sustain. Energy Rev.*, vol. 75, no. November 2016, pp. 293–312, Aug. 2017.
28. P. Prakash and D. K. Khatod, "Optimal sizing and siting techniques for distributed generation in distribution systems: A review," *Renew. Sustain. Energy Rev.*, vol. 57, pp. 111–130, May 2016.
29. F. S. Abu-Mouti and M. E. El-Hawary, "Optimal distributed generation allocation and sizing in distribution systems via artificial bee colony algorithm," *IEEE Trans. Power Deliv.*, vol. 26, no. 4, pp. 2090–2101, 2011.
30. T. Niknam, S. I. Taheri, J. Aghaei, S. Tabatabaei, and M. Nayeripour, "A modified honey bee mating optimization algorithm for multiobjective

placement of renewable energy resources," *Appl. Energy*, vol. 88, no. 12, pp. 4817–4830, Dec. 2011.
31. S. Sultana and P. K. Roy, "Krill herd algorithm for optimal location of distributed generator in radial distribution system," *Appl. Soft Comput. J.*, vol. 40, pp. 391–404, 2016.
32. N. Kanwar, N. Gupta, K. R. Niazi, A. Swarnkar, and R. C. Bansal, "Simultaneous allocation of distributed energy resource using improved particle swarm optimization," *Appl. Energy*, vol. 185, pp. 1684–1693, 2017.
33. R. Sanjay, T. Jayabarathi, T. Raghunathan, V. Ramesh, and N. Mithulananthan, "Optimal Allocation of Distributed Generation Using Hybrid Grey Wolf Optimizer," *IEEE Access*, vol. 5, pp. 14807–14818, 2017.
34. P. D. P. Reddy, V. C. V. Reddy, and T. G. Manohar, "Application of flower pollination algorithm for optimal placement and sizing of distributed generation in Distribution systems," *J. Electr. Syst. Inf. Technol.*, vol. 3, no. 1, pp. 14–22, May 2016.
35. A. Ulbig, T. S. Borsche, and G. Andersson, "Impact of Low Rotational Inertia on Power System Stability and Operation," *IFAC Proc. Vol.*, vol. 47, no. 3, pp. 7290–7297, 2014.
36. A. Adrees, P. N. Papadopoulos, and J. V Milanovi, "A Framework to Assess the Effect of Reduction in Inertia on System Frequency Response," *Power Energy Soc. Gen. Meet.*, pp. 1–5, 2016.
37. S. K. Pandey, S. R. Mohanty, and N. Kishor, "A literature survey on load-frequency control for conventional and distribution generation power systems," *Renew. Sustain. Energy Rev.*, vol. 25, pp. 318–334, Sep. 2013.
38. Y.-S. Kim, E.-S. Kim, and S.-I. Moon, "Frequency and Voltage Control Strategy of Standalone Microgrids With High Penetration of Intermittent Renewable Generation Systems," *IEEE Trans. Power Syst.*, vol. 31, no. 1, pp. 1–11, 2015.
39. J. Suh, D. H. Yoon, Y. S. Cho, and G. Jang, "Flexible Frequency Operation Strategy of Power System With High Renewable Penetration," *IEEE Trans. Sustain. Energy*, vol. 8, no. 1, pp. 192–199, 2017.
40. E. Rokrok, M. Shafie-khah, and J. P. S. Catalão, "Review of primary voltage and frequency control methods for inverter-based islanded microgrids with distributed generation," *Renew. Sustain. Energy Rev.*, vol. 82, no. March, pp. 3225–3235, 2018.
41. M. Abbes and M. Allagui, "Participation of PMSG-based wind farms to the grid ancillary services," *Electr. Power Syst. Res.*, vol. 136, pp. 201–211, 2016.
42. W. Bai, M. R. Abedi, and K. Y. Lee, "Distributed generation system control strategies with PV and fuel cell in microgrid operation," *Control Eng. Pract.*, vol. 53, pp. 184–193, 2016.
43. F. Díaz-González, M. Hau, A. Sumper, and O. Gomis-Bellmunt, "Participation of wind power plants in system frequency control: Review of grid code requirements and control methods," *Renew. Sustain. Energy Rev.*, vol. 34, pp. 551–564, 2014.

44. J. Morren, S. W. H. de Haan, W. L. Kling, and J. A. Ferreira, "Wind turbines emulating inertia and supporting primary frequency control," *IEEE Trans. Power Syst.*, vol. 21, no. 1, pp. 433–434, 2006.
45. S. Mishra, P. P. Zarina, and C. P. Sekhar, "A novel controller for frequency regulation in a hybrid system with high PV penetration," *IEEE Power Energy Soc. Gen. Meet.*, 2013.
46. D. Kim et al., "Virtual Inertial Control of a Wind Power Plant using the Frequency Deviation and the Maximum Rate of Change of Frequency Virtual Inertial Control of a Wind Power Plant using the Frequency Deviation and the Maximum Rate of Change of Frequency," vol. 8972, 2014.
47. P. P. Zarina, S. Mishra, and P. C. Sekhar, "Photovoltaic system based transient mitigation and frequency regulation," *2012 Annu. IEEE India Conf. INDICON 2012*, pp. 1245–1249, 2012.
48. P. P. Zarina, S. Mishra, and P. C. Sekhar, "Exploring frequency control capability of a PV system in a hybrid PV-rotating machine-without storage system," *Int. J. Electr. Power Energy Syst.*, vol. 60, pp. 258–267, 2014.
49. L. D. Watson and J. W. Kimball, "Frequency regulation of a microgrid using solar power," in *2011 Twenty-Sixth Annual IEEE Applied Power Electronics Conference and Exposition (APEC)*, 2011, pp. 321–326.
50. M. F. M. Arani and E. F. El-Saadany, "Implementing virtual inertia in DFIG-based wind power generation," *IEEE Trans. Power Syst.*, vol. 28, no. 2, pp. 1373–1384, 2013.
51. C. Pradhan, C. N. Bhende, and A. K. Samanta, "Adaptive virtual inertia-based frequency regulation in wind power systems," *Renew. Energy*, vol. 115, pp. 558–574, 2018.
52. J. Fang, H. Li, Y. Tang, and F. Blaabjerg, "Distributed Power System Virtual Inertia Implemented by Grid-Connected Power Converters," *IEEE Trans. Power Electron.*, vol. 8993, no. c, 2017.
53. J. Fang, X. Li, and Y. Tang, "Grid-connected power converters with distributed virtual power system inertia," in *2017 IEEE Energy Conversion Congress and Exposition (ECCE)*, 2017, vol. 2017-Janua, pp. 4267–4273.
54. W.-S. Im, C. Wang, W. Liu, L. Liu, and J.-M. Kim, "Distributed virtual inertia based control of multiple photovoltaic systems in autonomous microgrid," *IEEE/CAA J. Autom. Sin.*, vol. 4, no. 3, pp. 512–519, 2017.
55. E. Waffenschmidt and R. S. Y. Hui, "Virtual inertia with PV inverters using DC-link capacitors," pp. 1–10.
56. R. Bhatt and B. Chowdhury, "Grid frequency and voltage support using PV systems with energy storage," *NAPS 2011 - 43rd North Am. Power Symp.*, pp. 1–6, 2011.
57. M. Chamana and B. H. Chowdhury, "Droop-based control in a photovoltaic-centric microgrid with Battery Energy Storage," *45th North Am. Power Symp. NAPS 2013*, pp. 5–10, 2013.

58. L. Xin and W. Ning, "Study on the Control of Photovoltaic inverter based on Virtual Synchronous Generator," *IEEE 11th Conf. Ind. Electron. Appl.*, pp. 1612–1615, 2016.
59. A. Anzalchi, M. M. Pour, and A. Sarwat, "A combinatorial approach for addressing intermittency and providing inertial response in a grid-connected photovoltaic system," *IEEE Power Energy Soc. Gen. Meet.*, vol. 2016-Nov., pp. 1–5, 2016.
60. M. Datta, H. Ishikawa, H. Naitoh, and T. Senjyu, "Frequency control improvement in a PV-diesel hybrid power system with a virtual inertia controller," *Proc. 2012 7th IEEE Conf. Ind. Electron. Appl. ICIEA 2012*, pp. 1167–1172, 2012.
61. Yu J, Fang J, Tang Y. Inertia emulation by flywheel energy storage system for improved frequency regulation. In: 2018 IEEE 4th south power electron conf SPEC; 2018.
62. Kim YS, Kim ES, Moon S, "Frequency and voltage control strategy of stand-alone microgrids with high penetration of intermittent renewable generation systems." *IEEE Trans. Power Syst.*, 2016;31:718–28.
63. Fang J, Tang Y, Li H, Li X. A battery/ultracapacitor hybrid energy storage system for implementing the power management of virtual synchronous generators. *IEEE Trans Power Electron.*
64. Y. Hu, W. Wei, Y. Peng, and J. Lei, "Fuzzy virtual inertia control for virtual synchronous generator," in *Chinese Control Conference, CCC*, 2016, vol. 2016-Augus., pp. 8523–8527.

4

Energy Storage Systems for Electric Vehicles

M. Nandhini Gayathri

Department of EEE, School of Electrical and Electronics Engineering, SASTRA Deemed University, Thanjavur, India

Abstract

This chapter describes the growth of Electric Vehicles (EVs) and their energy storage system. The size, capacity and the cost are the primary factors used for the selection of EVs energy storage system. Thus, batteries used for the energy storage systems have been discussed in the chapter. The desirable characteristics of the energy storage system are enironmental, economic and user friendly. So the combination of various energy storage systems is suggested in EVs to present-day transportation. Apart from the selection of an energy storage system, another major part to enhance the EV is its charging. The fast charging schemes save battery charging time and reduce the battery size. The recent growth in power semiconductor, topology and intelligent charging control techniques reduce the expenditure of fast charging. In addition to the types of electric vehicles and classification of energy storage systems, other topics such as charging schemes, issues and challenges and recent advancements of the energy storage system of electric vehicle applications have also been discussed.

Keywords: Battery electric vehicle, hybrid electric vehicle, plugin hybrid EV, power electronics, battery, ultra-capacitors, flywheels, energy management

4.1 Introduction

Owing to escalating concerns over progressively worsening climatic conditions across the globe, and the steady depletion of non-renewable resources

---

Email: nandhini.gayathri@gmail.com

Sandeep Dhundhara and Yajvender Pal Verma (eds.) *Energy Storage for Modern Power System Operations*, (79–104) © 2021 Scrivener Publishing LLC

of energy and their environmental impact, it is important for the human community to shift to more responsible and more sustainable technologies. One of the most important future technologies is Electric Vehicles (EVs). It is slowly progressing towards reigning in all spheres of the transportation industry. In the overwhelming ecological crises [1], there is a need for efficient and reliable Electric vehicles. It can be manufactured by the conglomeration of various disciplines of science and engineering. EVs can inject the power back into the grid while not in operation.

A basic electric vehicle constitutes the following systems [2]: 1) The source of energy that powers the vehicle, 2) the power electronic converter circuitry, and 3) the electric motor system which converts the energy from the energy source to electric energy to drive the vehicle. The energy source for an electric vehicle [3] is obtained from the renewable sources of energy and hence the quantum of pollution associated with the operation of the EV is extremely minimal. In addition to the mechanical framework which forms the hardware, the EV also includes complex electric circuitry for energy transfer to various sections of the vehicle, and electronic circuitry for control, communication, and data transmission.

Although it does appear quite normal to view the Electric Vehicle technology as a modern and relatively new area of development, it is surprising to know that the first electric vehicle was designed in 1834. Until 1918, electric vehicles were very well in service and operated for public utility at modest costs. A survey taken in 1918 claimed that, of the 4,200 vehicles that operated on the road, a whopping 38% were electric vehicles, while 40% of vehicles were powered by steam, and 22% by gasoline [4]. As the fossil fuel-powered vehicles offered higher range and higher speed at comparatively lower costs, electric vehicles slowly began disappearing from the automobile market in over a decade, and so the technologies of internal combustion engine developed greatly compared to EV technologies. By 1933, the number of operating electric vehicles had declined to practically zero, and further contributions to improve the shortcomings of the EV technology also came to a screeching halt.

With increasing concerns around the deteriorating climatic situation and the huge contribution of the fossil fuel–powered vehicle sector to it, in the 1970s, the laboratories of the United States of America reinitiated active research towards addressing the worries in operation, and thus, there were better electric vehicles. This research was greatly facilitated by the extraordinary progress in the domains of microelectronics and power electronics over the four decades. The official order mandating the use of zero-emission vehicles [5] issued by the California Air Resources

Board in October 1990 made people and researchers alike understand the positive ecological impacts of EVs, and that the future [6] of the automobile industry would undeniably be the Electric Vehicle technology. With EV technology becoming one of the world's most attractive areas of research, contributions to the technology so far have improved EVs by remarkably reducing oil consumption, reducing the levels of carbon emissions, improving the safety, security, efficiency, compactness and reliability, while parallelly working towards reducing the costs involved to the lowest possible. The scope and demand for research in the domain of EV has not sufficed and is expected to attract more interest in the forthcoming decades as well.

For the EV technology to translate from paper to practice, i.e., operate on-road, the two most important concerns that need to be addressed are, 1) improving the range, 2) reducing the costs involved. To improve the range, the need is to build batteries of high energy density, and for better cost efficiency, the focus is on building simpler and better battery charging mechanisms, motors, speed control and other power electronic circuitry, and more efficient energy storage systems. The current average capacity of EVs can drive the vehicle to a speed of 95.6km/h in 10 seconds with 62kWh [7], and the figures are as good as conventional vehicles that consume 31kWh to accelerate to attain a range of 100 miles [8]. One of the most important elements that need to be considered in the design of EV is its specific energy capacity since it defines the range of the vehicle. For a Hybrid Electric Vehicle (HEV), the specific power consideration is prioritized over specific energy considerations since it helps in maintaining the steady and efficient performance of the vehicle during operation, especially when high power is required during acceleration, the ascent of slopes and regenerative braking.

As electric vehicles are powered by non-conventional, less polluting sources of renewable energy which require energy storage systems (ESS), it calls for improving the ESSs by providing more safety of operation and compactness and improving the efficiency and reliability while reducing costs. Similarly, with advancements, as hybrid electric vehicles supported by efficient power electronics increase, the challenge of optimizing the power required to run them arises. This chapter presents a comprehensive review of construction, control and conversion of the various technologies of ESSs and their characteristics, modes of power delivery, with their respective advantages and limitations for EV applications. The various classification of ESS based on formation and composition, their efficiencies over the entire operation period and evaluation processes are also discussed. Although the present ESS technologies can be used with an EV, the pinnacle of use of ESS technologies for efficient storage of energy has not yet been attained.

4.2 Energy Storage Systems for Electric Vehicle

Integrating an ESS with the grid helps in providing more reliable, flexible, and steady power to consumers during a power outage or congestion, due to unforeseen events. It also stores power during off-peak operation period, and supplies to the grid when demand increases, thereby lessening the electricity cost. Hence, ESSs play a very important role in the upcoming micro-grid and smart grid technologies. Thus, EVs with ESS are also a potential source of energy to the grid, and since EVs run on non-polluting sources of energy, it becomes important to develop the capabilities of the ESSs to their fullest. To meet the requirements of the electric vehicle, we may also use a combination of ESSs to increase the reliability and the duration of discharge [9–17]. The drive train architecture of the EVs is presented in Figure 4.1.

i. Types of Energy Storage System
The energy storage system is classified into six types based on the energy formations and composition materials as shown in Figure 4.2. The further classifications of various ESS are shown in Figure 4.3 to 4.8. The ESS systems classification is based on the nature of energy utilization [12, 18–26].

4.3 Types of Electric Vehicles

EVs are available in three different combinations. An EV with an internal combustion engine (ICE) is known as a hybrid electric vehicle (HEV), an EV with a battery alone configuration is known as electric vehicle (BEV) and a hybrid electric vehicle with external charging facility is known as a plug-in hybrid electric vehicle (PHEV) by utilizing the electric propulsion system. The use of batteries in EV is advantageous compared to conventional technologies. EVs produce less noise in operation, are more pliable in removing flue gas pollutants and the major advantage is that the exploitation cost of EV is three times lower. The disadvantages of battery systems are their heavy mass, the high expenditure, and higher size which pose constraints on range and performance, along with implications due

Energy Storage Systems for Electric Vehicles 83

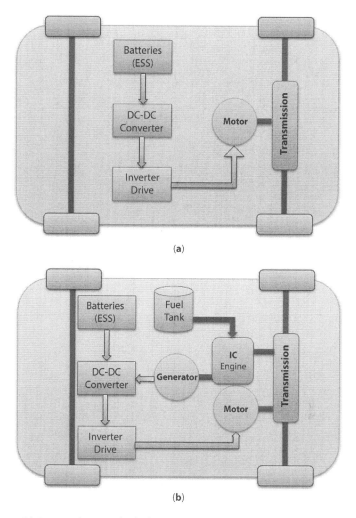

Figure 4.1 (a) Battery electric vehicle drive system (b) Series-parallel full HEV.

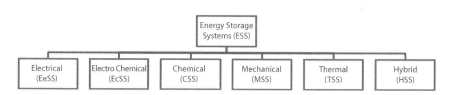

Figure 4.2 Types of Energy Storage Systems (ESS).

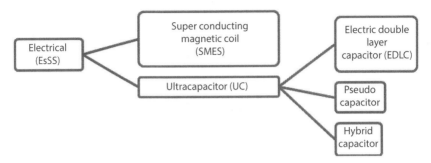

Figure 4.3 Classification of Electrical Storage Systems (ESS).

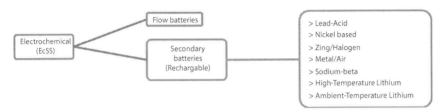

Figure 4.4 Classification of Electro-Chemical Energy Storage Systems (EcSS).

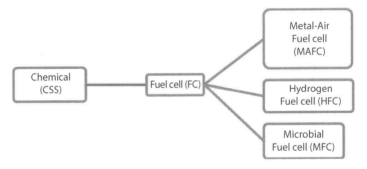

Figure 4.5 Classification of Chemical Energy Storage Systems (CSS).

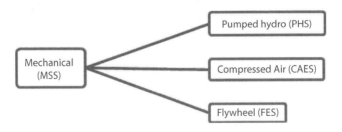

Figure 4.6 Classification of Mechanical Energy Storage Systems (MSS).

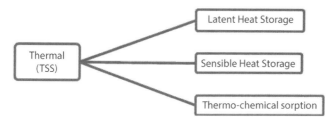

Figure 4.7 Classification of Thermal Energy Storage Systems (TSS).

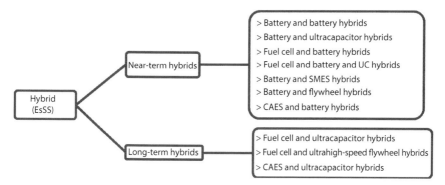

Figure 4.8 Classification of Hybrid Energy Storage Systems (HSS).

to varied climatic conditions. Hence, EVs require more work to make them more reliable and efficient [27–38].

4.3.1 Battery Electric Vehicle (BEV)

The power to the BEV is supplied from a high-capacity battery. BEV may also have its power source from the grid. However, since the batteries are majorly depended upon, the major elements that demand attention are climatic condition, cost and energy density, which are the basis of classification.

According to [36], Batteries greatly help to reduce the polluting emissions and they can also be integrated with the grid to provide clean power back into the grid derived from various renewable sources. The first EV was launched as tricycle power by a battery in 1834. Figure 4.9 shows the basic configuration of battery electrical vehicle.

The range of BEVs are 100 to 400 km, and it is dependent on the capacity of the battery used. The cell configuration and capacity determine the time needed to charge the battery for a cycle and its life. It is affected by the ambient temperature too. To increase the range of EVs, HEV, PHEV, etc., can be used.

Figure 4.9 Battery electric vehicle.

4.3.2 Hybrid Electric Vehicle (HEV)

When more than one energy source is used to power the vehicle, it is known as a hybrid vehicle. In a hybrid electric vehicle, an electric motor and an internal combustion (IC) engine are combined. It is the most proven and established procedure in practice. The engine-driven from the gasoline is one type of HEV, and the other is a bi-directional energy storage system. Regenerative braking is implemented in EV to improve efficiency. The traditional IC engine causes harmful exhaust and wastage of fuel during varied operation cycles. HEV works based on the transmission of electric energy through the motor, thereby overcoming the shortcomings. Another advantage of this is that the vehicle can be driven on electric power with its peak range when there is no fuel. The HEV is broadly classified into three categories based on the construction.

i. Series Hybrid
In a series hybrid, the range is extended by ICE. The configuration of series hybrid electric vehicle has been presented in Figure 4.10. An electric generator produces electricity to charge a battery and run the motor. In this system, the generator acts as the supply for both of the batteries and the motor.

The battery pack and the motor are bigger while the IC engine has a smaller size. The absence of a mechanical interface between the IC engine and transmission causes maximum efficiency.

ii. Parallel Hybrid
In the parallel hybrid type, the mechanical transmission is made by a parallel connection between the IC engine and electric motor.

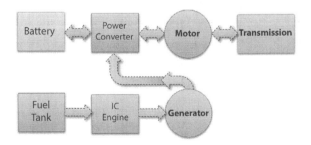

Figure 4.10 Series hybrid electric vehicle.

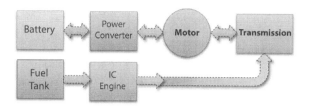

Figure 4.11 Parallel hybrid electric vehicle.

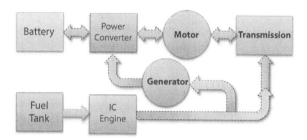

Figure 4.12 Combined hybrid electric vehicle.

In usual cases, the engine runs as the primary means and backup or torque power booster is the electric motor. This needs only lighter and compact batteries. The regenerative braking supports the battery recharging at cruising mode, As shown in Figure 4.11, there is a fixed mechanical interface from the wheels to the EV motor. So it is not possible to charge when the vehicle is not in motion.

iii. Combined Hybrid

The integration of the above two types is called as a combined hybrid or series-parallel or complex/power split system. It supplies power to the vehicle in both ways, mechanically as well as electrically. Both mechanical and electrical power is interconnected, as shown in Figure 4.12. The real-time implementation of this system is complex and ten times costlier than BEVs.

4.3.3 Plug-In Hybrid Electric Vehicles (PHEV)

Electric power or gasoline is used to energize the PHEV. In PHEV, the medium-capacity batteries are used. It provides better range when compared with other EVs. But they are more expensive than BEVs and not entirely eco-friendly. It restricts the kilometre range and maximum speed when compared with other gasoline-powered vehicles.

4.4 Review of Energy Storage Systems for Electric Vehicle Applications

4.4.1 Key Attributes of Battery Technologies

Several attributes are involved in the classification of different battery energy storage systems especially in the transportation applications [39]. The key attributes of battery are energy and power density, working temperature, charge retainment capability, cell voltage, cost per kilowatt-hour (kWh), battery safety and battery recyclability [40].

4.4.2 Widely Used Battery Technologies

The main technologies are Li-Ion, LA, Ni-Cd, NiMH and NaNiCl2 and Flow batteries [41–55].

i. Lead-Acid battery chemistry
It is the most popular rechargeable battery type, in which the cathode and anode are made of PbO_2 Pb, respectively. The sulfuric acid is used as the electrolyte medium. The quick response time, low self-discharge rates, maximum cycle efficiencies and minimum capital cost are the benefit of this battery.

ii. Lithium-Ion battery chemistry
In this battery cathode and anode are made up of Lithium metal oxide ($LiCoO_2$, $LiMO_2$, etc.) and Graphitic carbon, respectively. The non-aqueous organic liquid (such as $LiClO_4$) is employed as the electrolyte. Response time is less, and they have higher power per volume. The cycle efficiencies of Li-Ion batteries are also high [58–60, 101]. The structure of the Li-Ion battery is shown in Figure 4.13.

iii. NiMH battery chemistry
The anode and cathode of this type of battery are made up of Nickel hydroxide and multi-components of vanadium, titanium, nickel, and some other metals. In the past two decades, there has been a great development in NiMH battery technology. The energy capacity increased three times with a 10 times increase in the specific power. High operating voltages, excellent energy and power per a unit volume, ability to tolerate overcharge

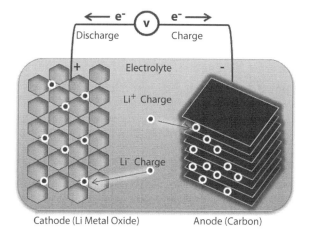

Figure 4.13 Li-Ion battery chemistry.

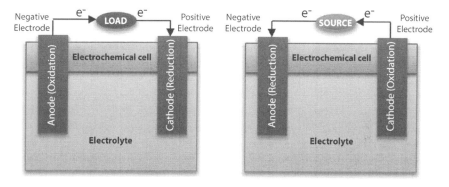

Figure 4.14 NiMH battery chemistry.

and over-discharge and commendable thermal properties are the major advantages. The NiMH battery chemistry is illustrated in Figure 4.14.

iv. Ni-Cd battery chemistry
The anode and cathode are made up of Nickel hydroxide and metallic cadmium, respectively. The aqueous alkali solution is used as the electrolyte. The robust structure results in high reliability and low maintenance. The toxic heavy metals are a major drawback and cause a hazard. This can be highlighted as one weakness of NiCd batteries.

v. NaNiCl2 battery chemistry
The NaNiCl2 batteries have a liquid Na electrode & β"-alumina solid electrolyte. It is also known as ZEBRA (Zeolite Battery Research Africa) batteries [74, 75].

The anode is made from a secondary electrolyte made of molten sodium tetra chloroaluminate (NaAlCl4). Insoluble nickel chloride is the active material. The Sodium Nickel chloride has the following good properties: high specific density, temperature agnostic, long life and long storage life, no memory effect, maintenance-free and zero ambient emission and complete recyclability. It is used in electric and plug-in hybrid vehicles. The structure of NaNiCl2 battery is shown in Figure 4.15.

vi. Flow batteries

The most common type of flow batteries is Vanadium sulphate – Vanadium oxide sulphate [76]. The oxidation of V2+ to V3+ takes place at one electrode of the battery, while a reduction of V5+ to V4+ takes place in the other. These reactions take place at carbon or graphite electrodes without the actual reaction. The structure of Flow battery chemistry is shown in

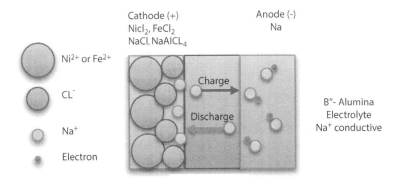

Figure 4.15 NaNiCl2 battery chemistry.

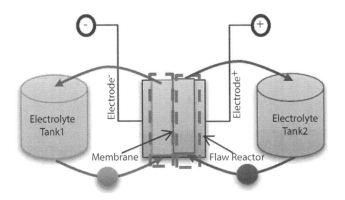

Figure 4.16 Flow battery chemistry.

Table 4.1 Performance comparison of different types of batteries.

Battery type	Lead-acid	Ni-Cd	Ni-MH	Zn-Br	Li-ion
Energy Density (Wh/kg)	30-50	45-80	60-120	35-54	110-160
Power Density (W/kg)	180	150	250	1000	1800
Nominal Voltage	2V	1.25V	1.25V	1.67V	3.6V
Operating Temperature	20-60°	40-60°	20-60°	20-60°	20-60°
Cycle Life	200-300	1500	300-500	>2000	500-1000
Charge Efficiency (%)	79	-	-	-	100
Energy Efficiency (%)	70	60-90	75	80	80
Overcharge Tolerance	High	Moderate	Low	High	Very low
Self-discharge	Low	Moderate	High	Low	Very low
Thermal Stability	Least stable	Least stable	Least stable	Least stable	Most stable

Figure 4.16. The present research is focused on the increase in energy and power capacity, etc.

The performance comparison of different battery types [100] is presented in Table 4.1. The high power and energy density of Lithium-ion battery are more suitable for EV applications.

4.4.3 Alternate Energy Storage Solutions

In automotive applications, FCs, UCs, FESS are used in addition to battery energy storages [56–68].

i. Ultra-Capacitor Technologies
In UCs, the high permittivity dielectric with the high surface area is used to store considerable energy. Carbon and metal fibre composites, aerogel carbon, particulate carbon with a binder, conducting polymer films (doped) on carbon cloth, coatings of mixed metal oxide on metal foil are the present technologies employed in UCs.

ii. Fuel Cell Technology
In fuel technology, the hydrogen is used as the source fuel cells. It completely replaces the IC engine in automotive applications. Hydrogen is produced by the process of electrolysis of water. The process is known as the Hydrogen Economy.

iii. Flywheel Energy Storage System
Use of Flywheel Energy Storage System in the automotive sector is a new approach. A flywheel is a giant rotating disk and stores kinetic energy. A coupling setup of motor/generator and flywheel helps to perform the conversion between kinetic energy and electrical energy. The electric motor supplies kinetic energy and increases the rotational speed of the flywheel. The generator receives stored energy from the flywheel and drives the load. The high power and energy density and the infinite number of charging-discharging are the major advantages. A partial vacuum setup is required in FESS implementation to reduce wind-based losses. To reduce the frictional losses active magnetic bearings are used. The FESS is high-cost system used in large vehicles that need a battery system of a bigger size.

iv. Lithium-Ion Capacitor
The Lithium-Ion Capacitor is a new hybrid innovative technology and lies between Lithium-Ion batteries and supercapacitors. The construction is similar to a Lithium-Ion battery but is pre-doped with Lithium.

The cathode is activated carbon, similar to an Electrochemical Double Layer Capacitor (EDLC). It is used in electric vehicle quick-charging stations.

v. Hybrid Energy Storage System
The combination of at least two energy storage systems is known as a hybrid energy storage system. It helps to obtain the mutual benefit and better characteristics of the energy storage system. It provides a better capability to handle electric power peaks and the regenerative braking operations.

4.5 Electric Vehicle Charging Schemes

The charger is the essential component of BEV to charge it. The process of charging involves advanced control techniques for regulation of current and voltage. The EV mount charger is known as built-in-charger. In the charging station, it is called a stand-alone charger. Battery's safety, durability, and performance are determined by the charging and discharging process [69]. The various charging methodologies are constant current (CC), constant voltage (CV), and the combination of constant voltage and constant current (CVCC) technique. For regenerative braking to happen, batteries require random charging. The various levels of charging of EVs are explained below [70–72].

i. Slow Charging
This is used in a residential outlet (230 V-AC). An on-board charger is present in all EVs, with the power of about 2 kW. It is the simplest battery charging method used for charging which takes over a single night, and the vehicle can travel four miles per hour.

ii. Semi-Fast Charging
These charging stations can have a charging power of five times higher than the slow charging. It provides about 16 miles per hour charging of travel with a 3.4-kW onboard charger.

iii. Fast Charging
This is used for DC fast charging (DCFC), and it offers 350 km range for half an hour charging. Due to the excess power level, the charger is present off-board. Here, the EV's traction batteries are fed with high power DC through the charging inlet present on the EV. The charging infrastructure planning scenario in India [102] is shown in Figure 4.17.

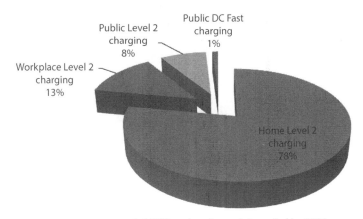

Figure 4.17 EV Charging infrastructure forecast for 2030.

4.6 Issues and Challenges of ESSs in EV Applications

The current evolution of ESSs is satisfactory for storing and controlling EV resources. It also decreases oil production, CO_2 emissions, and GHGs. With technical advances, ESSs are increasingly maturing [73, 74]. Still, it suffers due to other reasons such as raw material handling, energy management, power converters interface, sizing, security measures and the cost [75, 76]. The key issues related to energy storage systems are listed below.

- The availability and supply of good quality raw materials for manufacturing ESSs.
- Suitable power converter with an advanced control algorithm.
- Utilization of renewable energy sources.
- Large-scale implementation.
- Pollution-free electrochemical batteries.
- Safety and reliability.

4.7 Recent Advancements in the Storage Technologies of EVs

The performance of ESS is measured in terms of efficiency and life cycle. The amount of power delivery and its efficiency varies for different ESSs. The EV requires 10kW to 1000kW power for a few hours of operation [77–80].

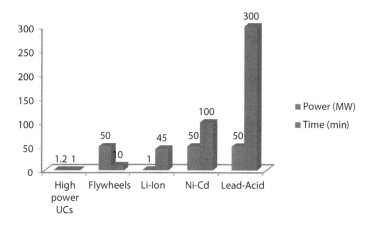

Figure 4.18 Energy storage technologies for EV and Transportation applications.

Figure 4.18 shows the time taken by ESS to deliver the power for EVs and other transportation applications. The secondary battery, ultra-capacitors and high-speed flywheel are used to supply EVs.

The life cycle of each ESS also varies for different ESS. The Superconducting magnetic ESSs have a maximum efficiency of 97% and a long life cycle at 80% DOD. Similarly, ultra-capacitors and flywheels also have maximum efficiencies and long life cycles. The batteries have 70-85% of efficiency with a life cycle of 2000-4500 at 80% DOD. Whereas, the fuel cells have low efficiency and high life cycles [81–87]. The various ESSs are combined to standardize the next-generation EV applications in transportation. Efficient energy storage is obtained by hybrid ESS in which batteries and fuel cells are preferred as primary ESS and ultra-capacitors, fuel cells are selected as auxiliary ESS [88, 89].

Even though the present scenario supports the development of ESS to a satisfactory level for EV applications, there is still a need for attention on the raw material availability and its appropriate disposal, energy management, power converter interface, sizing and safety measures and cost. A few recommendations are mentioned here to improve the efficiency and battery time-span in the present and future research [90–95].

The recycling, refurbishing and reuse of used ESS materials are recommended to avail cost-effective, efficient and environmentally friendly solution.

The low loss high-efficiency power electronic converters are preferred along with intelligent control algorithm to improve the efficiency, battery storage time-span and avoid battery abuse.

An effective and intelligent EMS suggested handling all available energy sources to increase the battery storage time-span as well as energy.

Large-scale storage is preferred to achieve good efficiency at a good price as there is a trade-off between efficiency and cost [96–99].

4.8 Factors, Challenges and Problems in Sustainable Electric Vehicle

This section presents the review of the hurdles and various elements that affect the market growth and diffusion of EVs.

i. Economic Challenge
Cost-effectiveness is one of the major factors which determine usage of EVs across the globe. Any system in practice should be economically recommended. The cost of EV implementation is divided into three parts: EV, battery and infrastructure.

The capital investment of EV is very high when compared to ICE vehicles and results in a high selling price. The quality of the battery increases the costs. The charging cost is also based on the quality of the battery. Thus, current research is focused on developing technologies with lower cost component. Public charging infrastructures should be developed for the penetration of EVs into markets and to overcome the range anxiety of EV users. In general, if the number of EVs is higher, the charging demand is higher, and the profitability and usage become higher.

ii. Technological Challenge
The major technological challenge is the energy storage system. EVs can operate on power obtained from batteries, which require recharging from one or more temporary energy resource. Emerging EV technology utilises all of the possible energy resources effectively. Due to high material cost, maintenance cost as well as labour cost, the price of ESS gets increased. The better selection of battery and intelligent charging scheme with good infrastructure help to reduce the ESS price.

iii. Social and Environmental Challenge
An affordable EV price, government incentives, and reliable technology will promote the usage of EVs in coming years. The charging of EV also produces greenhouse gas emissions but with lesser amount. Increased

usage of EVs, which reduces greenhouse emissions by replacing ICE vehicles and using renewable energy sources, will have a huge impact on it.

4.9 Conclusions and Recommendations

The EVs market is continuously growing and is expected to occupy future transportation completely. More investment is being made in the research and development of EVs technologies, thus accelerating global EV transitions. The International Energy Agency (IEA) has reported a global stock increment of EVs from 3.7 million to 13 million from 2017 to 2020 and eventually it is expected to reach 130 million in 2030. Sales of EVs are estimated to have an average growth of 24% throughout this period. Sales would increase from 1.4 million to 4 million EVs from 2017 to 2020 and this figure is expected to reach 21.5 million sales by 2030.

EV technologies, especially the battery density and capacity, as well as its lifecycle, can be further advanced. The new technologies and mass production will result in further price reduction of EVs. The operation and maintenance cost is reduced by increasing the number of public charging infrastructures, providing incentives and subsidies.

By following the above recommendation, the price of EVs could be reduced and the EV customer experience can be enhanced. Therefore, the number of EVs will increase in the coming years leading to a significant reduction in carbon emissions, thereby ensuring cleaner and better environmental conditions in the future.

References

1. Chan, C., 1999. Engineering philosophy of electric vehicles. In: *IEEE International Electric Machines and Rives Conference*, Seattle, 1999.
2. Chan, C., 1993. An overview of electric vehicle technology. *Proceedings of the IEEE* 81 (9), 1202e1213.
3. Chan, C., Chau, K., 1997. An overview of power electronics in electric vehicles. *IEEE Transactions on Industrial Electronics* 44, 3e13.
4. Rajashekara, K., 2013. Present status, and future trends in electric vehicle propulsion technologies. *IEEE Journal of Emerging and Selected Topics in Power Electronics* 1, 3e10.
5. Zhang, Q., Li, C., Wu, Y., 2017. Analysis of research and development trend of the battery technology in electric vehicle with the perspective of patent. *Energy Procedia* 105, 4274e4280.

6. Wang, Y.W., 2007. An optimal location choice model for recreation-oriented scooter recharge stations. *Transportation Research Part D: Transport and Environment* 12 (3), 231e237.
7. Zhang, Q., Li, C., Wu, Y., 2017. Analysis of research and development trend of the battery technology in electric vehicle with the perspective of patent. *Energy Procedia* 105, 4274e4280.
8. Frieske, B., Kloetzke, M., Mauser, F., 2013. Trends in vehicle concept and key technology development for hybrid and battery electric vehicles. In: 2013 *World Electric Vehicle Symposium and Exhibition,* Barcelona, 2013.
9. United States (US) Climate Action Report (CAR). U.S. Department of State. Available from: (http://www.state.gov/documents/organization/219038.pdf); 2014 [30.6.2015].
10. Olivier JGJ, Janssens-Maenhout G, Muntean M, Peters JAHW. Trends in global CO2 emissions: 2014 Report. PBL Netherlands Environmental Assessment Agency, The Hague; 2014.
11. Hacker F, Harthan R, Matthes F, Zimmer W. Environmental impacts and impact on the electricity market of a large scale introduction of electric cars in Europe critical review of literature. *ETC/ACC technical paper.* 2009; 4: 56–90.
12. Electrical Energy Storage. White paper. International Electrotechnical Commission. (IEC), Geneva, Switzerland; 2011.
13. Hardin D. Smart grid and dynamic power management. Energy management systems, GiridharKini (Ed.), InTech. Available from: (http://www.intechopen.com/books/energymanagement-systems/smart-grid-and-dynamic-powermanagement); 2011. [30.6.2015].
14. *Smart Grid Solutions for Power Infrastructure & Industrial Energy Systems. Smart grid solutions guide 2014.* USA: Texas Instruments (TI); 2014.
15. Han Y, Xu L. A survey of the smart grid technologies: background, motivation and practical applications. *Przegląd Elektrotech (Electr Rev)* 2011:87.
16. Fang X, Misra S, Xue GL, Yang DJ. Smart grid - The new and improved power grid: a survey. *IEEE Commun Surv Tutor* 2012;14(4):944–80.
17. Tie SF, Tan CW. A review of energy sources and energy management system in electric vehicles. *Renew Sustain Energy Rev* 2013;20:82–102.
18. Chau KT, Wong YS, Chan CC. An overview of energy sources for electric vehicles. *Energy Convers Manag* 1999;40:1021–39.
19. Anderman M. Status and trends in the HEV/PHEC/EV battery industry. Colorado, USA: Rocky Mountain Institute; 2008.
20. Hannan MA, Azidin FA, Mohamed A. Hybrid electric vehicles and their challenges: a review. *Renew Sustain Energy Rev* 2014;29:135–50.
21. Lukic SM, Cao J, Bansal RC, Rodriguez F, Emadi A. Energy storage systems for automotive applications. *IEEE Trans Ind Electron* 2008;55(6): 2258–67.
22. Azidin FA, Hannan MA, Mohamed A. Renewable energy technologies and hybrid electric vehicle challenges. *Prz Elektrotech* 2013;89(8):150–6.

23. Madanipour V, Montazeri-Gh M, Mahmoodi-k M. Multi-objective component sizing of plug-in hybrid electric vehicle for optimal energy management. *Clean Technol Environ Policy* 2016;18(4):1189–202.
24. Duvall M, Alexander M. Batteries for electric drive vehicles - status 2005: performance, durability, and cost of advanced batteries for electric, hybrid electric, and plug-in hybrid electric vehicles. *Electr Power Res Inst* 2005 [November].
25. Dhameja S. Electric *vehicle battery systems*. USA: Newnes publication, Butterworth–Heinemann; 2002.
26. Vazquez S, Lukic SM, Galvan E, Franquelo LG, Carrasco JM. Energy storage systems for transport and grid applications. *IEEE Trans Ind Electron* 2010;57(12).
27. Singh, B., Jain, P., Mittal, A., et al., 2006. Direct torque control: a practical approach to electric vehicle. In: *2006 IEEE Power India Conference*, New Delhi, 2006.
28. Koniak, M., Czerepicki, A., 2017. Selection of the battery pack parameters for an electric vehicle based on performance requirements. In: *IOP Conference Series: Materials Science and Engineering*, Pitesti, 2017.
29. Nitti, M., Pilloni, V., Colistra, G., et al., 2015. The virtual object as a major element of the internet of things: a survey. *IEEE Communications Surveys & Tutorials* 18 (2), 1228e1240.
30. De Luca, S., Di Pace, R., Marano, V., 2015. Modelling the adoption intention and installation choice of an automotive aftermarket mild-solar-hybridization kit. *Transportation Research Part C: Emerging Technologies* 56, 426e445.
31. Pinsky, N.R., Argueta, J.C., Knipe, T.J., et al., 2000. Fast charge of lead-acid batteries at the SCE EV Tech Center. In: *The Fifteenth Annual Battery Conference on Applications and Advances*, Long Beach, 2000.
32. Emadi, A., 2005. *Handbook of Automotive Power Electronics and Motor Drives*. CRC Press, Boca Raton, Florida.
33. Kebriaei, M., Niasar, A.H., Asaei, B., 2015. Hybrid electric vehicles: an overview. In: *2015 International Conference on Connected Vehicles, and Expo (ICCVE)*, Shenzhen, 2015.
34. Thompson, T.M., King, C.W., Allen, D.T., et al., 2011. Air quality impacts of plug-in hybrid electric vehicles in Texas: evaluating three battery charging scenarios. *Environmental Research Letters* 6, 024004.
35. Shen, C., Shan, P., Gao, T., 2011. A comprehensive overview of hybrid electric vehicles. *International Journal of Vehicular Technology* 2011 (S1), 1e7.
36. Pollet, B.G., Staffell, I., Shang, J.L., 2012. Current status of hybrid, battery and fuel cell electric vehicles: from electrochemistry to market prospects. *Electrochimica Acta* 84, 235e249.
37. Beresteanu, A., Li, S., 2011. Gasoline prices, government support, and the demand for hybrid vehicles in the United States. *International Economic Review* 52, 161e182.

38. Wirasingha, S.G., Emadi, A., 2009. Pihef: plug-in hybrid electric factor. *IEEE Transactions on Vehicular Technology* 60 (3), 1279e1284.
39. Daoud MI, Massoud AM, Abdel-Khalik AS, Elserougi A, Ahmed S. A flywheel energy storage system for fault ride through support of grid-connected VSC HVDC-based offshore wind farms. *IEEE Trans Power* Syst 2016;31(3):1671–80.
40. Hiroshima N, Hatta H, Koyama M, Yoshimura J, Nagura Y, Goto K, Kogo Y. Spin M.A. Hannan et al. *Renewable and Sustainable Energy Reviews* 69 (2017) 771–78;9 Spin test of three-dimensional composite rotor for flywheel energy storage system. Compos Struct 2016;136:626–34.
41. Li L, Wu Z, Yuan S, Zhang XB. Advances and challenges for flexible energy storage and conversion devices and systems. *Energy Environ Sci* 2014;7(7):2101–22.
42. Linden D, Reddy TB. *Handbook of batteries*, 3rd ed. New York: McGraw-Hill, Inc; 2001
43. Ibrahim H, Ilinca A, Perron J. Energy storage systems - characteristics and comparisons. *Renew Sustain Energy Rev* 2008;12:1221–50.
44. Lee JH, Yoon CS, Hwang JY, Kim SJ, Maglia F, Lamp P, Myung ST, Sun YK. Highenergy- density lithium-ion battery using a carbon-nanotube-Si composite anode and a compositionally graded Li[Ni0.85Co0.05Mn0.10]O-2 cathode. *Energy Environ Sci* 2016;9(6):2152–8.
45. Renn G, Ma G, Cong N. Review of electrical energy storage system for vehicular applications. *Renew Sustain Energy Rev* 2015;41:225–36.
46. Hadjipaschalis I, Poullikkas A, Efthimiou V. Overview of current and future energy storage technologies for electric power applications. *Renew Sustain Energy Rev* 2009;13:1513–22.
47. Li JC, Zhang QL, Xiao XC, Cheng YT, Liang CD, Dudney NJ. Unravelling the impact of reaction paths on mechanical degradation of intercalation cathodes for lithium-ion batteries. *J Am Chem Soc* 2015;137(43):13732–5.
48. Li GS, Lu XC, Kim JY, Meinhardt KD, Chang HJ, Canfield NL, Sprenkle VL. Advanced intermediate temperature sodium-nickel chloride batteries with ultrahigh energy density. *Nat Commun* 2016;7:10683.
49. Zhou ZB, Benbouzid M, Charpentier JF, Scuiller F, Tang TH. A review of energy storage technologies for marine current energy systems. *Renew Sustain Energy Rev* 2013;18:390–400.
50. Khaligh A, Li Z. Battery, ultracapacitor, fuel cell, and hybrid energy storage systems forelectric, hybrid electric, fuel cell, and plug-in hybrid electric vehicles: state of the art. *IEEE Trans Veh Technol* 2010;59(6):2806–14.
51. Divya KC, Østergaard J. Battery energy storage technology for power systems- an overview. *Electr Power Syst Res* 2009;79:511–20.
52. Garcia-Plaza M, Serrano-Jimenez D, Carrasco JEG, Alonso-Martinez J. A Ni-Cd battery model considering state of charge and hysteresis effects. *J Power Sources* 2015;275:595–604.

53. Akhil AA, Huff G, Currier AB, Kaun BC, Rastler DM, Chen SB, Cotter AL, Bradshaw DT, Gauntlett WD DOE/EPRI 2013 Electricity Storage Handbook in Collaboration with NRECA, SANDIA REPORT, SAND2013-5131, Sandia National Laboratories; July 2013(www.sandia.gov/ess).
54. Chen HS, Cong TN, Yang W, Tan CQ, Li YL, Ding YL. Progress in electrical energy storage system: a critical review. *Prog Nat Sci* 2009;19:291–312.
55. Li YG, Gong M, Liang YY, Feng J, Kim JE, Wang HL, Hong GS, Zhang B, Dai HJ. Advanced zinc-air batteries based on high-performance hybrid electrocatalysts. *Nat Commun* 2013;4:1805.
56. Sapkota P, Kim H. Zinc–air fuel cell, a potential candidate for alternative energy. *J Ind Eng Chem* 2009;15(4):445–50.
57. Li GS, Lu XC, Kim JY, Viswanathan VV, Meinhardt KD, Engelhard MH, Sprenkle VL. An advanced Na-FeCl2 zebra battery for stationary energy storage application. *Adv Energy Mater* 2015;5(12):1500357.
58. Er S, Suh C, Marshak MP, Aspuru-Guzik A. Computational design of molecules for an all-quinone redox flow battery. *Chem Sci* 2015;6(2):885–93.
59. Zhao Y, Ding Y, Li YT, Peng LL, Byon HR, Goodenough JB, Yu GH. A chemistry and material perspective on lithium redox flow batteries towards high-density electrical energy storage. *Chem Soc Rev* 2015;44(22):7968–96.
60. Gong K, Fang QR, Gu S, Li SFY, Yan YS. Nonaqueous redox-flow batteries: organic solvents, supporting electrolytes, and redox pairs. *Energy Environ Sci* 2015;8(12):3515–30.
61. Wei XL, Xu W, Huang JH, Zhang L, Walter E, Lawrence C, Vijayakumar M, Henderson WA, Liu TB, Cosimbescu L, Li B, Sprenkle V, Wang W. Radical compatibility with nonaqueous electrolytes and its impact on an all-organic redox flow battery. *Angew Chem-Int Ed* 2015;54(30):8684–7.
62. Liu H, Jiang J. Flywheel energy storage – an upswing technology for energy sustainability. *Energy Build* 2007;39(5):599–604.
63. Bolund B, Bernhoff H, Leijon M. Flywheel energy and power storage systems. *Renew Sustain Energy Rev* 2007;11(2):235–58.
64. van Berkel K, Rullens S, Hofman T, Vroemen B, Steinbuch M. Topology and flywheel size optimization for mechanical hybrid powertrains. *IEEE Trans Veh Technol* 2014;63(9):4192–20565.
65. Xu Y, Pi HW, Ren TQ, Yang Y, Ding HF, Peng T, Li L. Design of a multipulse highmagnetic- field system based on flywheel energy storage. *IEEE Trans Appl Supercond* 2016;26(4):5207005
66. Xu KX, Wu DJ, Jiao YL, Zheng MH. A fully superconducting bearing system for flywheel applications. *Supercond Sci Technol* 2016;29(6):064001.
67. Ogata M, Matsue H, Yamashita T, Hasegawa H, Nagashima K, Maeda T, Matsuoka T, Mukoyama S, Shimizu H, Horiuchi S. Test equipment for a flywheel energy storage system using a magnetic bearing composed of superconducting coils and superconducting bulks. *Supercond Sci Technol* 2016;29(5):054002.

68. Yuan Y, Sun YK, Huang YH. Design and analysis of bearingless flywheel motor specially for flywheel energy storage. *Electron Lett* 2016;52(1):66-7.
69. Ahmadian, A., Sedghi, M., Aliakbar-Golkar, M., 2015. Stochastic modeling of plug-in electric vehicles load demand in residential grids considering non-linear battery charge characteristic. In: *20th Conference on Electrical Power Distribution Networks*, Zahedan, 2015.
70. Dost, P., Spichartz, P., Sourkounis, C., 2015. Charging behavior of users utilizing battery electric vehicles and extended-range electric vehicles within the scope of a field test. In: *2015 International Conference on Renewable Energy Research and Applications (ICRERA)*, Palermo, 2015.
71. Rahman, I., Vasant, P.M., Singh, B.S.M., et al., 2016. On the performance of accelerated particle swarm optimization for charging plug-in hybrid electric vehicles. *Alexandria Engineering Journal* 55, 419e426.
72. Perry, M.L., Fuller, T.F., 2002. A historical perspective of fuel cell technology in the 20th century. *Journal of the Electrochemical Society* 149, S59eS67.
73. Bromaghim G, Serfass J, Serfass P, Wagner E. Hydrogen and fuel cells. The U.S. market report; 2010.
74. Semadeni M. Energy storage as an essential part of sustainable energy systems: a review on applied energy storage technologies [May]. CEPE, ETH Zentrum, Zürich; 2003.
75. Chen W, Ådnanses AK, Hansen JF, Lindtjørn JO, Tang T. Super-capacitors based hybrid converter in marine electric propulsion system. *Proc IEEE XIX Int Electr Mach Conf*, Rome 2010:1-6.
76. Liu S, Sun S, You X. Inorganic nanostructured materials for high performance electrochemical supercapacitors. *Nanoscale* 2014;6(4):2037-45.
77. Dubal DP, Ayyad O, Ruiz V, Gomez-Romero P. Hybrid energy storage: the merging of battery and supercapacitor chemistries. *Chem Soc Rev* 2015;44(7):1777-90.
78. Keil P, Englberger M, Jossen A. hybrid energy storage systems for electric vehicles: an experimental analysis of performance improvements at subzero temperatures. *IEEE Trans Veh Technol* 2016;65(3):998-1006.
79. Sun B, Dragicevic T, Freijedo FD, Vasquez JC, Guerrero JM. A control algorithm for electric vehicle fast charging stations equipped with flywheel energy storage systems. *IEEE Trans Power Electron* 2016;31(9):6674-6685.
80. Yoo H, Sul SK, Park Y, Jeong J. System integration and power flow management for a series hybrid electric vehicle using supercapacitors and batteries. *IEEE Trans Ind Appl* 2008;44(1):108-14.
81. Shuai L, Corzine KA, Ferdowsi M. A new battery/ultracapacitor energy storage system design and its motor drive integration for hybrid electric vehicles. *IEEE Trans Veh Technol* 2007;56(4):1516-23.
82. Henson W. Optimal battery/ultracapacitor storage combination. *J Power Sources* 2008;79(1):417-23.

83. Liu W, Kang D, Zhang CN, Peng GH, Yang XF, Wang SY. Design of a high-T-C superconductive maglev flywheel system at 100-kW level. *IEEE Trans Appl Supercond* 2016;26(4):5700805.
84. Machado F, Trovao JPF, Antunes CH. Effectiveness of supercapacitors in pure electric vehicles using a hybrid metaheuristic approach. *IEEE Trans Veh Technol* 2016;65(1):29–36.
85. Anno T, Koizumi H. Double-input bidirectional DC/DC converter using cellvoltage equalizer with flyback transformer. *IEEE Trans Power Electron* 2015;30(6):2923–34.
86. Nahavandi A, Hagh MT, Sharifian MBB, Danyali S. A nonisolated multiinput multioutput DC-DC boost converter for electric vehicle applications. *IEEE Trans Power Electron* 2015;30(4):1818–35.
87. Yang G, Dubus P, Sadarnac D. Double-phase high-efficiency, wide load range high- voltage/low-voltage LLC DC/DC converter for electric/hybrid vehicles. *IEEE Trans Power Electron* 2015;30(4):1876–86.
88. Ostadi A, Kazerani M. A comparative analysis of optimal sizing of battery-only, ultracapacitor-only, and battery-ultracapacitor hybrid energy storage systems for a city bus. *IEEE Trans Veh Technol* 2015;64(10):4449–60.
89. Soylu S. Electric vehicles—modelling and simulations. DC/DC converters for electric vehicles, 13. Rijeka, Croatia: *InTech*; 2011. p. 478.
90. Govindaraj A. Design and characterization of various circuit topologies for battery/ultracapacitor hybrid energy storage systems. North Carolina: NC State University; 2010.
91. Budzianowski WM. Negative carbon intensity of renewable energy technologies involving biomass or carbon dioxide as inputs. *Renew Sustain Energy Rev* 2012;16(9):6507–21.
92. Budzianowski WM. Sustainable biogas energy in Poland: prospects and challenges. *Renew Sustain Energy Rev* 2012;16(1):342–9.
93. Sulaiman N, Hannan MA, Mohamed A, Majlan EH, Daud WRW. A review on energy management system for fuel cell hybrid electric vehicle: issues and challenges. *Renew Sustain Energy Rev* 2015;52:802–14.
94. Zakerin B, Syri S. Electrical energy storage systems: a comparative life cycle cost analysis. *Renew Sustain Energy Rev* 2015;42:569–96.
95. Li L, Dunn JB, Zhang XX, Gaines L, Chen RJ, Wu F, Amine K. Recovery of metals from spent lithium-ion batteries with organic acids as leaching reagents and environmental assessment. *J Power Sources*, 7 2013;233:180–9.
96. Gaines L. The future of automotive lithium-ion battery recycling: Charting a sustainable course. *Sustain Mater Technol* 12 2014(1–2):2–7.
97. Dunn JB, Gaines L, Sullivan J, Wang MQ. The impact of recycling on cradle-to-gate energy consumption and greenhouse gas emissions of automotive lithiumion batteries. *J Chem Educ, Environ Sci Technol Am Chem Soc Publ* 11 2012;46(22):12704–10.

98. Bhagat K, Saha SK. Numerical analysis of latent heat thermal energy storage using encapsulated phase change material for solar thermal power plant. *Renew Energy* 2016;95:323–36.
99. Zhao C, Yin H, Ma CB. Quantitative evaluation of LiFePO4 battery cycle lifeimprovement using ultracapacitors. *IEEE Trans Power Electron* 2016;31(6):3989–93.
100. M. A. Hannan, M. M. Hoque, A. Hussain, Y. Yusof and P. J. Ker, "State-of-the-Art and Energy Management System of Lithium-Ion Batteries in Electric Vehicle Applications: Issues and Recommendations," in *IEEE Access*, vol. 6, pp. 19362-19378, 2018, doi: 10.1109/ACCESS.2018.2817655.
101. Weidong Chen, Jun Liang, Zhaohua Yang, Gen Li, "A Review of Lithium-Ion Battery for Electric Vehicle Applications and Beyond," *Energy Procedia*, Volume 158, 2019, pp. 4363-4368, ISSN 1876-6102, https://doi.org/10.1016/j.egypro.2019.01.783.
102. M. Bilal and M. Rizwan, "Electric vehicles in a smart grid: a comprehensive survey on optimal location of charging station," in *IET Smart Grid*, vol. 3, no. 3, pp. 267-279, 6 2020, doi: 10.1049/iet-stg.2019.0220.

5

Fast-Acting Electrical Energy Storage Systems for Frequency Regulation

Mandeep Sharma[1], Sandeep Dhundhara[2], Yogendra Arya[3]*
and Maninder Kaur[4]

[1]*Department of Electrical Engineering, Baba Hira Singh Bhattal Institute of Engineering and Technology, Lehragaga, Punjab, India*
[2]*Department of Basic Engg., College of Agricultural Engg. and Tech., CCS Haryana Agricultural University, Hisar, India*
[3]*Department of Electrical Engineering, J.C. Bose University of Science and Technology, YMCA, Faridabad, Haryana, India*
[4]*Dr. S.S.B. University Institute of Chemical Engineering & Technology, Panjab University, Chandigarh, India*

Abstract

Energy storage plays a very important role in maintaining frequency regulation in interconnected electric power systems having diverse generating units. Enormous frequency deviations appear if load frequency control (LFC) capacity is incompetent to compensate for the imbalances of generation and demand. In this type of scenario, fast-acting energy storage systems are one of the best options to tackle such types of challenges while maintaining energy generation-demand balance. The main emphasis of this study is to analyze the application of fast-responding energy storage systems to regulate frequency in modern power systems having multiple generating units. The impact of electrical energy storage systems like capacitive energy storage (CES), and superconducting magnetic energy storage (SMES) has been analysed with different operating situations of the traditional and deregulated power systems. These storage technologies show promising results in the investigated power systems.

Keywords: Deregulated power system, capacitive energy storage (CES), superconducting magnetic energy storage (SMES), wind generator, photovoltaic (PV) generator, frequency control, optimization algorithm

---

*Corresponding author: mr.y.arya@gmail.com

Sandeep Dhundhara and Yajvender Pal Verma (eds.) *Energy Storage for Modern Power System Operations*, (105–142) © 2021 Scrivener Publishing LLC

5.1 Introduction

Uninterrupted power supply to consumers with high efficiency of electric power systems (EPS) and reduction in greenhouse gas emission is the foremost challenge for power engineers in today's fast-progressing society [1]. Inefficient operation of EPS may cause wastages of energy, blackouts, and system failures that could harm electrical types of equipment and cause substantial financial loss to the supplier and the end users. Therefore, because of the environmental worries, there has been tremendous pressure worldwide to limit the utilization of hydrocarbons for electric power generation. These concerns have led to a larger integration of renewables in the electric grid. But, the stochastic nature of these resources presents issues with grid reliability, stability, and power quality [2].

In this specific circumstance, the utilization of electrical energy storage (EES) systems has been proposed to build efficiency, stability, and reliability of the EPS by providing energy savings and ancillary power supply. Energy storage is viewed as the empowering innovation for various applications that lessens the greenhouse gas outflows and improve system-level efficiency [3]. Further, larger integration of renewables in the electric grid demands weeks to months of energy storage in the future (up to GWh to TWh scale) to mitigate their intermittent nature [4].

Presently, an incredible effort is being made by researchers, policymakers, and the industry everywhere in the world towards the field of energy storage systems, demonstrating developments and progress in system-level and component design in power electronic interfaces, modeling, and control, and practical integration and implementation issues [5]. Various interesting issues in this field are related to applications like aircraft applications, electric vehicles, grid integration, etc., and developments like new and hybrid energy storage systems, efficiency, the life span of energy storage systems, and power electronics interface for energy storage resources (requirements, topologies, control, and design), etc. [3].

5.1.1 Significance of Fast-Acting Electrical Energy Storage (EES) System in Frequency Regulation

The EES system has the following key roles in the power system:

1. Increases system reliability during sudden failures.
2. By storing energy at off-peak rates, it helps to reduce the electricity supply costs.

3. Preserve and improve system stability, reliability, and power quality (voltage and frequency).
4. Compensate for the intermittent behaviour of renewable energy resources.

In general terms, EES systems are applied in various applications as a buffer that permits the whole system to operate more efficiently, stably, and reliably. Various energy storage systems used for the above-mentioned roles are electrochemical type capacitors, batteries, fuel cells, superconductors, compressed air, flywheels, etc. Among various storage systems, current applications of storage systems are mostly centered around capacitive energy storage (CES) and superconducting magnetic energy storage (SMES). The electrical energy is stored in the form of electrostatic and electromagnetic energy in CES and SMES, respectively. The other chief EES system is ultra-capacitor (UC) or double-layer capacitor, generally called super-capacitor. Here, the main medium to store electrical energy is electrostatic, whereas a superconducting conductor is used to store the electrical energy in an electromagnetic storage system [6]. In today's fast-progressing society, utilization and storage of the most usable form of energy, i.e., electrical generated by renewable as well as sustainable protocol are the crucial challenges. These challenges prompted advancement in electrochemical energy storage devices such as batteries, fuel cells, and super-capacitors. Even though batteries and fuel cells have higher energy density, unreasonable installation price, massiveness, short lifetime and low power capabilities are significant restrictions to date. Regardless, super-capacitors have some intrinsic characteristics, for example, lightweight, fast charging rate, simple movability, low maintenance cost, and high life span with the probability for high specific and energy densities [7]. These alluring highlights have prompted scientific challenges that have advanced the desire for supplanting traditional gigantic and low-lasting batteries. A brief introduction to the electrostatic and electromagnetic EES systems follows below.

5.1.2 Capacitive Energy Storage (CES)

Capacitors directly store the electricity in electrostatics form. The key constituents of a CES unit are power conversion system (PCS) and some protective devices, and a capacitor (a super-capacitor or a cryogenic hyper-capacitor) to store the energy [8]. During the storage process, super-capacitors work on the same principle as capacitors. However, super-capacitors have higher energy density due to the replacement of insulating material of conventional capacitor by an electrolyte ionic conductor which provides a very large specific surface for ion movement along a conducting electrode [9]. During ordinary operational conditions, the CES unit uses its capacitor to store the energy in the form of static charge and instantaneously discharges its stored energy into the grid for the duration of any unexpected load disruption. The schematic illustration of the CES unit is provided in Figure 5.1.

Numerous key features of CES, such as rapid response time, speedy charging and discharging rate without effecting the overall efficacy for numerous cycles, extended service life, extraordinary power density, and having an ample ability to compensate for frequent power requests, make CES suitable for different applications of power system operation [10].

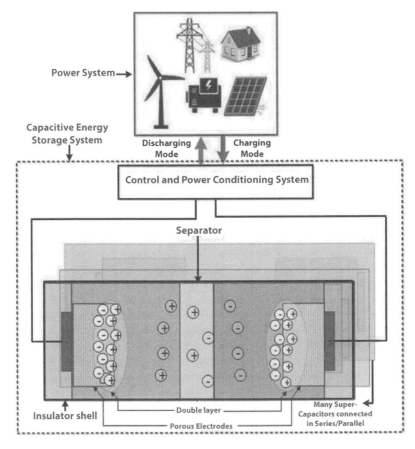

Figure 5.1 Schematic diagram of CES [6].

In comparison to other energy storage units, CES having a self-protecting system is thus expected to display fewer failure rates and extraordinary fault-tolerant competency with more durability. CES with the high energy efficiency of approximately 95%, have 40-50% less initial cost to store the energy than that of the SMES unit with the same power rating [8]. The mere losses which take place are the energy losses due to the PCS, self-discharge, and internal leakage [11]. However, the high cost of the super-capacitors which is estimated at five times that lead-acid battery cost is one of the major drawbacks of this technology [9].

A CES is best suitable for the power system ancillary applications such as power quality support, frequency regulation, and load leveling. Various global researchers are on the way to improving the characteristics of CES such as higher operating temperature limits, energy density, longer lifespan of the super-capacitor, and reducing super-capacitor cost. Recently, scientists at the University of Central Florida developed a super-capacitor prototype by integrating graphene as electrodes, which can achieve a specific energy density of more than 50 Wh/kg and has 20 times longer life than Li-ion batteries. In other research, the conductivity and surface area of the super-capacitor is enhanced by using graphene [6].

5.1.2.1 Basic Configuration of CES

The key components of a CES unit are a super-capacitor/cryogenic hyper-capacitor to store the energy, PCS, and some protective devices [8]. The schematic of a CES system is shown in Figure 5.1. A CES unit stores energy in its capacitor plates in the form of static charge and instantly dispenses its stored energy into the grid to keep the generation and demand in balance. Thus, additionally, it supports the governor and other control schemes to bring the power system into a new equilibrium condition. The CES unit again retains its initial voltage value across the capacitor plates by exploiting leftover energy within the system. The circuit diagram of the CES unit is displayed in Figure 5.2.

The PCS comprises the rectifier and inverter units to convert the electric energy from AC-DC and vice versa. It can likewise give an electrical interface between the capacitor and the power framework. The resistor 'R_1' associated in parallel over the capacitor is the lumped equivalent resistance demonstrating the leakage and dielectric losses of the capacitor bank. To minimize the harmonics in capacitor output voltage and on the AC bus, two bridges arrangement is ideal. The bypass thyristors offer a path for current I_d if the converter failure occurred. The DC breaker enables the

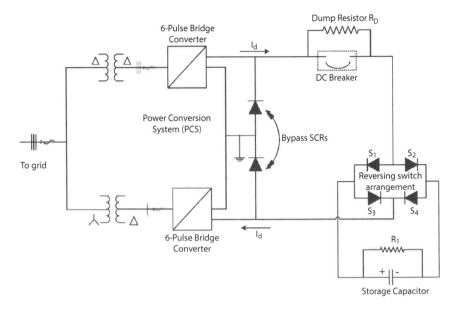

Figure 5.2 Circuit diagram of CES unit.

current I_d to be redirected towards the energy dump resistor R_D in case of the converter failure. Considering negligible losses, the E_d is stated as follows [12]:

$$E_d = 2E_{do}\cos\alpha - 2I_d R_c \quad (5.1)$$

where E_d denotes the DC bridge voltage (kV) applied to the capacitor, E_{do} represents the maximum open-circuit voltage of each bridge, α indicates the firing angle (degree), I_d represents the current (kA) following through the capacitor whereas R_c is the equivalent commutating resistance (ohm) of the CES unit [12]. The voltage E_d over the capacitor can be regulated between its extreme positive and extreme negative values, by varying the relative phase angle α between within a range from 0° to 180°. The energy stored (E_s) at the capacitor plates at any instant of time can be expressed as follows:

$$E_s = 0.5 C E_d^2 \text{ MJ} \quad (5.2)$$

where C is the capacitance of the capacitor in farads [13]. To activate the conduction of each silicon controlled rectifier (SCR) at a set time, the voltage pulses are generated accordingly from the firing circuits. This sequence keeps up a steady average voltage over the capacitor. The careful planning to generate firing pulses concerning the phase of 50/60 Hz AC voltage decides the average DC bridge voltage over the capacitor. Meanwhile, the bridges all the time keep up unidirectional current and α exclusively characterize E_d for positive and negative values and control the power P_d in the capacitor in both direction and magnitude. In this way, by continually governing the firing angle α, even with no switching operation, magnitude control, as well as the reversibility of the power flow is accomplished. The converter firing angle is constrained by an algorithm designed as per the utility needs; however, essentially the control circuit reacts to a request signal for a specific power level, either negative or positive. At that point, depending on the voltage over the capacitor, a firing angle is determined and transmitted into the firing circuit [12, 13].

Since the change in direction of capacitor current is not happening through the bridge converter (inverter/rectifier), the reversing switch mechanism facilitates this change during discharging (for peak load period) and charging (for rated load period). In the charging mode, switches S_2, S_3 are OFF and S_1, S_4 are ON. During discharging mode, switches S_1, S_4 are OFF and S_2, S_3 are ON [12].

The main goal of the capacitor is to keep maximum permissible energy release equal to the maximum permissible energy absorption. This makes the CES unit extremely successful in alleviating the alternations generated by unexpected load perturbations. If E_{do} indicates the established estimation of voltage, E_{dmax} and E_{dmin} indicate the upper and lower bounds of voltage, respectively. In mathematical form,

$$\frac{1}{2}CE_{dmax}^2 - \frac{1}{2}CE_{do}^2 = \frac{1}{2}CE_{do}^2 - \frac{1}{2}CE_{dmin}^2 \quad (5.3)$$

Hence,

$$E_{do} = \frac{[E_{dmax}^2 + E_{dmin}^2]^{\frac{1}{2}}}{2} \quad (5.4)$$

The $P_{do} = E_{do}I_{do}$ and $P_d = E_dI_d$ represents the initial and instantaneous power flow into the capacitor, respectively. I_{do} and E_{do} are the magnitudes of current and voltage before the load disruption [12]. The alteration in the power output of the CES during a load disturbance condition can be described as follows:

$$P_{do} + \Delta P_d = (E_{do} + \Delta E_d)(I_{do} + \Delta I_d) \quad (5.5)$$

The incremental power change in the capacitor can be represented as follows:

$$\Delta P_d = (\Delta E_d I_{do} + E_{do} \Delta I_d) \tag{5.6}$$

The term $E_{do}I_{do}$ is neglected since E_{do} is zero ($E_{do} = 0$) during storage mode to keep the rated voltage at a constant value.

5.1.2.2 CES Control Logic

The frequency deviation (ΔF) from the power system is exploited to control the CES current I_d. The incremental change in current is stated as follows [12, 14]:

$$I_{di} = \left[\frac{1}{1+sT_{CES}}\right] K_{CES}\; F_{Area\text{-}i}(s) \tag{5.7}$$

where, ΔI_{di} is the incremental change in CES unit current (kA), K_{CES} denotes the gain constant of the control loop (kA/Hz), and T_{CES} signifies the time constant of the CES representing converter time delay (s), s is the Laplace operator $\left(\dfrac{d}{dt}\right)$ and 'i' indicates the area (e.g. i = 1,2). Figure 5.3 represents the transfer function model for CES unit as a frequency stabilizer [28].

The incremental variation in the power of the CES including phase compensation blocks is represented as follows:

$$\Delta P_{CES} = \left[\frac{K_{CES}}{1+sT_{CES}}\right]\left[\frac{1+sT_{C1}}{1+sT_{C2}}\right]\left[\frac{1+sT_{C3}}{1+sT_{C4}}\right] \Delta F_{Area\text{-}i}(s) \tag{5.8}$$

where, i = 1,2, and T_1, T_2, T_3, and T_4 are the time constants of two-stage phase compensation blocks [15]. CES unit delivers the requisite amount of power analogous to the signal i.e. ΔF which acts as a control signal for the CES unit. To limit the storage and release capacity of CES for the system with 2000 MVA base capacity, a limiter of values up to 0.01 pu (ΔP_{max}) and

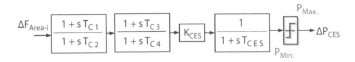

Figure 5.3 Linearized transfer function model of a CES as a frequency stabilizer.

−0.01 pu (ΔP_{min}) is added at the output of the CES as presented in Figure 5.3 [16]. A detailed explanation of the power modulation and configuration of CES units has been given in [17, 18].

5.1.3 Superconducting Magnetic Energy Storage (SMES)

SMES has the potential of recharging in minutes and discharging megawatts of power within a fraction of a cycle to fulfil any abrupt power demand of the grid. SMES can replicate the charge and discharge process thousands of times without any deprivation of the magnet. The recharging time mainly depends upon the system capacity and can be enhanced according to the system necessities for a specific application [18]. SMES is an electromagnetic EES system that stores electrical energy in a magnetic field produced by the movement of DC current in a superconducting coil. The superconducting coil contains niobium-titanate (NbTi) filaments that operate at superconducting critical temperature (−270°C) and have nearly zero internal resistance [9]. The main components of a SMES system comprise a superconducting coil made of an alloy, called 'niobium-titanium', a cryogenic refrigerator to maintain the operating temperature at a superconducting state, a PCS that controls the charging/discharging process, and a control mechanism that operates the whole system according to the specific application [19]. High efficiency (up to 95%), good power density, high depth of discharge (DoD), long cycle life, high reliability, and quick response time are the main advantages of SMES system which make it suitable for several applications of power system such as load levelling, frequency regulation, and power quality support [9]. However, high system cost, low storage capacity, short discharge duration, interference to nearby communication lines, and low energy density are the major obstacles in the wide implementation of SMES systems in power system applications [6, 19].

Technological advancement and developments in the low-cost high-temperature superconductors such as BSCCO (bismuth-strontium-calcium-copper oxide), MgB_2 (magnesium diboride), and YBCCO (yttrium-barium-calcium-copper oxide) coated conductors can bring a substantial decrease in the conductor as well as refrigeration system cost for the SMES unit in future [6].

5.1.3.1 Constructional and Working Details of SMES

In SMES units, the coil of niobium-titanium is kept at 4.2 K using liquid helium or nitrogen, the temperature essential to convert it into a

114 Energy Storage for Modern Power System Operations

superconductor. In the state of superconducting the coil has zero electrical resistance and thus can carry huge currents for long periods with a very small loss. A typical SMES unit comprises three parts: a power conditioning system cryogenically cooled refrigerator and superconducting coil. The current will not fall off after the superconducting coil is once charged. Thus, the magnetic energy can be stored for an indefinite time and the required stored energy can be delivered back into the grid by discharging the coil. The conversion of AC to DC power and vice versa is achieved by inverter/rectifier with 2-3% energy loss (in each direction) in the power conditioning system. SMES units are highly efficient with the least amount of electricity loss in the energy storage procedure in contrast to other approaches to energy storage. SMES systems have high round trip efficiency which is more than 95%. SMES unit is currently used for applications where short-duration energy storage is required because of energy requirements for refrigeration and high-cost superconducting wire and. Hence, SMES is most usually dedicated to refining power quality.

5.1.3.2 Basic Configuration of SMES

Figure 5.4 shows the schematic diagram of SMES unit having a thyristor control [20]. SMES unit comprises the converter and DC superconducting

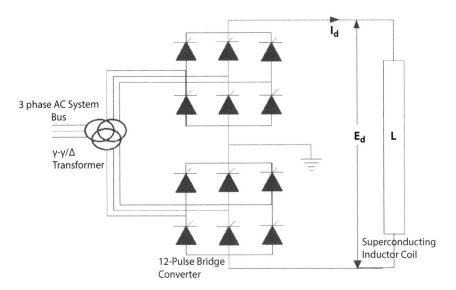

Figure 5.4 The schematic diagram of SMES unit.

coil that is connected by Y-Y/Y-D transformer. The DC voltage Ed appearing across the inductor controlled by the converter firing angle to vary it continuously within a certain range of positive and negative values.

The inductor by applying a small positive voltage is charged to its rated current I_{do} at the start. When the current attains its rated value, it is upheld at this value by dropping the voltage across the inductor to zero as the coil is superconducting [21]. Neglecting the losses in the converter and the transformer, the DC voltage can be calculated as in (5.9):

$$E_d = 2V_{do} \cos\alpha - 2I_d R_c \qquad (5.9)$$

where E_d and V_{do} are DC voltage applied across the inductor and the maximum circuit bridge voltage, respectively (kV), I_d is the current flowing through the inductor (KA), α is firing angle (degree), and R_c is equivalent commutating resistance (Ω). To control the charging and discharge mode of SMES, commutation angle α is changed:

In the case of α < 90°, the converter work in charging mode (converter mode), and for α >90°, the converter work in discharging mode (inverter mode).

In the LFC process, the dc voltage E_d across the superconducting inductor is continuously controlled depending on the sensed area control error (ACE) signal. In the literature, it is taken to be dependent on the sensed ΔF signal or ACE defined as $B_i \Delta F_i + \Delta Ptie_i$ (i=1,…,n) [21]. The inductor is first charged to its rated current, I_{do} by applying a little positive voltage. Once the current has attained the rated value, it is held constant by reducing voltage ideally to zero since the coil is superconducting. A very small voltage may be required to overcome the commutating resistance. The energy stored at any instant can be given as follows [20]:

$$W_L = \frac{L * I_d^2}{2} \text{ MJ} \qquad (5.10)$$

where w_L = Energy stored, MJ, I_d = Current flow in the coil, A, and L = Inductance of SMES, H.

5.1.3.3 SMES Block Diagram Presentation

Figure 5.5 shows the transfer function model (TFM) of SMES as frequency stabilizer, modeled as the second-order lead-lag compensator. The frequency deviation (ΔF) from the power system is used to control the DC voltage E_d across the superconducting inductor. To limit the storage and

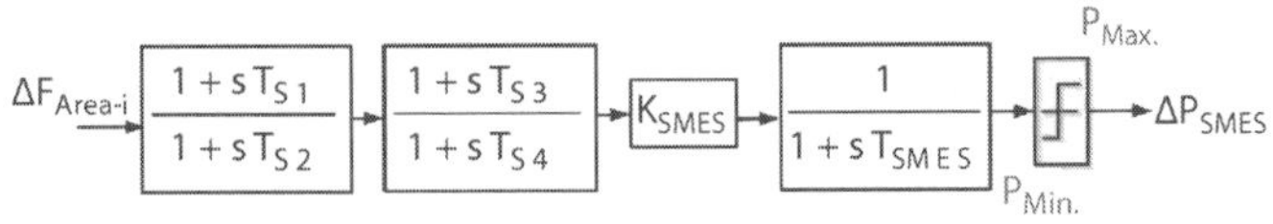

Figure 5.5 Linearized model of a SMES as a frequency stabilizer.

release capacity of SMES for the system with 2000 MVA base capacity, a limiter of values up to 0.01 pu (ΔP_{Max}) and -0.01 pu (ΔP_{Min}) is added at the output of the SMES as presented in Figure 5.5 [16]. The incremental variation in the power of the SMES including phase compensation blocks is represented as follows:

$$\Delta P_{SMES} = \left[\frac{K_{SMES}}{1+sT_{SMES}}\right]\left[\frac{1+sT_{S1}}{1+sT_{S2}}\right]\left[\frac{1+sT_{S3}}{1+sT_{S4}}\right]\Delta F_{Area-i}(s) \quad (5.11)$$

5.1.3.4 Benefits Over Other Energy Storage Methods

There are several advantages of SMES units over other electrical energy storage devices such as:

1. The most significant favourable position of SMES is that the time delay while charging and discharging is very short.
2. The power loss is less compared to other storage systems because in superconducting mode electric currents encounter almost zero resistance.
3. SMES possesses high reliability as no moving part exists in SMES unit.
4. SMES provides increased efficiency, extending operability and low maintenance of generating units on account of its ability to absorb the fluctuations and ramp at exceptionally rapid rates.
5. SMES unit, if deliberately positioned, can concede the requirement for new transmission by loading already existed transmission frameworks during off-peak periods.
6. SMES unit offers additional back-up thus increasing the overall capacity of the system that was earlier required only during peak periods.
7. SMES provides clean and efficient storage of electrical energy from renewable and conventional units and may offer some emission credit.

5.1.4 Advantages of CES Over SMES [22, 23]

The benefits of CES over SMES are summarized as follows:

1. CES system does not require any superconductive temperature during operation. However, SMES require cryogenic refrigerators for liquid helium (4.2 K) which results in a high operating cost of SMES compared to CES. The self-discharge losses and internal leakage currents are negligible in the CES unit, which makes it more efficient than a SMES system.
2. CES has a lower weight and relatively higher energy density compared to SMES units of the same power rating.
3. The capacity of any CES unit can be upgraded by integrating extra capacitor modules into the system, while the upgradation of SMES coil is not possible in this manner.
4. The problem of stray magnetic fields like in SMES is not present in the CES unit.
5. The coil of SMES unit develops strong internal Lorentz forces during discharge due to the interactions between the currents and the magnetic field, whereas such large forces are absent in the CES unit. So, the absence of force-related technical problems and failures makes the CES system more suitable compared to SMES unit.
6. The SMES can provide high power rates but it is associated with high voltages. To draw energy from the circulating current, voltage is generated across the superconducting coil by switching. For large power rates requiring fast discharges, may involve very large voltages. This increases the risks of inter-turn voltage breakdown due to high voltage gradients between the coil turns. Whenever a capacitor in a bank suffers a voltage failure, the resulting large momentary currents burn out the current path instantly. This provides CES a self-protecting property that is lacking in the other types of EES systems. Hence, CES is generally expected to have exceptionally low failure rates, to be fault-tolerant and robust. Also, no CES failures due to thermal shocks have ever been observed.
7. The initial cost involved in CES is lower. Preliminary estimates show that the initial cost of a CES is 40-50% less than that for SMES of the same capacity. In light of this, CES systems can provide an alternative for refining the

interconnected power system performance; while for storage purposes, either super-capacitors or cryogenic hyper-capacitors can be used in CES.

5.2 Case Study to Investigate the Impact of CES and SMES in Modern Power System

5.2.1 Literature Review

As a consequence of intense interest in renewable energy, recently enormous consideration has been paid to the use of EES systems in power systems. Also, excessive energy consumption creates a need for EES that can be used later as per need. The significant trends driving the market of EES involve the following:

- The rise in adoption of EES in the sector of transportation.
- Software integration enhances EES based management.
- Increase in renewable energy adoption.

Further, EES devices with and without coordination of flexible AC transmission systems (FACTS) devices can be applied in the power system (PS) for LFC application effectively. As EES provides storage facilities for short and excessive power demand thus it provides frequency regulation in addition to inertial support delivered by the rotating mass in the PS. EES system uses various energy storage units such as SMES, redox flow batteries (RFB), pumped storage, and CES for frequency regulation applications. The illustration of a commonly used scheme for LFC in a PS using the EES system is presented in Figure 5.6.

EES units are very effective in coping with the uncertain frequency oscillation due to their quick response and precise control. They curtail the grid frequency variations by compensating for inadequate power and

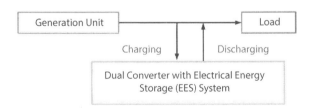

Figure 5.6 General illustration of the EES system in LFC.

Table 5.1 Literature review summary of several types of EES devices used for LFC in PS.

Ref. no.	Author name/year	Number of areas/ sources	Generation source type	System type	Controller type	Optimization techniques	FACTS/EES/other devices
[4]	Dhundhara and Verma/2020	2/Multi	Thermal-Hydro-Gas-DFIG based WTU	RPS	PI controller	MFO	RFB
[24]	Nosratabadi, et al./2019	3/Multi	Thermal-Hydro-Wind-Diesel-Gas	RPS	Predictive Functional Modified PID controller	Grasshopper optimization algorithm	RFB
[25]	Prakash et al./2019	2/Multi	SPV-Wind – Diesel-AE-FC	RPS	2-DOF-PI-FOPDN controller	VPL algorithm	HVDC tie-link
[26]	Sharma et al./2019	3/Single	Thermal	TPS	ANFIS	--	TCPS and SMES
[27]	Haroun and Yin-Ya/2019	2/Multi	Thermal-Hydro-Gas	TPS	FOPID controller integrated with an FO fuzzy PID controller	Craziness-based PSO	TCSC
[28]	Dhundhara and Verma/2018	2/Multi	Thermal-Hydro-Gas	RPS	PI controller	SCA	TCPS and CES

(Continued)

Table 5.1 Literature review summary of several types of EES devices used for LFC in PS. (*Continued*)

Ref. no.	Author name/year	Number of areas/ sources	Generation source type	System type	Controller type	Optimization techniques	FACTS/EES/other devices
[29]	Sharma et al./2018	6/Single	Thermal-Hydro-Gas-Nuclear-Wind-Diesel	TPS	PID controller	MVO	IPFC and RFB
[30]	Tasnin and Saikia/2018	2/Multi	GTPP-Thermal-Hydro-Gas-Wind-SP	TPS	FOPI-FOPID controller	SCA	Battery, Flywheel, CES, SMES, UC, and RFB
[31]	Morsali et al./2018	2/Multi	Thermal-Hydro-Gas	RPS	I controller	IPSO and MGSO	SSSC
[32]	Rajbongshi and Saikia/2017	3/Multi	SPPP-Thermal-Diesel-Wind	TPS	FOIDF controller	LSA	TCSC, SSSC and IPFC, SMES
[33]	Sudha and Santhi/2012	2/Single	Thermal	TPS	Type-2 fuzzy	----	SMES
[34]	Pappachen and Fathima/2019	3/Multi	Thermal-Hydro	RPS	PI controller	GA	SMES

Abbreviations: RPS - Restructured power system, **TPS** - Traditional power system, **DFIG** - Doubly fed induction generator, **WTU** - Wind turbine unit, **MFO** - Moth flame optimization, **SPV** - Solar photovoltaic, **VPL** - Volleyball premier league, **AE** - Aqua electrolyzer, **FC** - Fuel cell, **TCPS** - Thyristor controlled phase shifter, **TCSC** - Thyristor controlled series compensator, **SCA** - Sine cosine algorithm, **MVO** - Multi-verse optimizer, **GTPP** - geothermal power plant, **UC** - Ultra-capacitor, **RFB** - Redox flow battery, **SSSC** - Static synchronous series compensator, **IPSO** - improved particle swarm optimization, **MGSO** - Modified group search optimization, **STPP** - Solar thermal power plant, **FOIDF** - Fractional integral derivative controller with filter, **LSA** - Lightning search algorithm, **IPFC** - Interline power flow controller, **GA** - Genetic algorithm.

engrossing undue power during dynamic conditions, thus advancing the power quality of PS. Several studies have been performed which show the impact of EES devices in order to enhance the PS performances in terms of smooth frequency control as presented in Table 5.1. For minimal load perturbation even with efficient control executions, still, frequency fluctuations and tie-line power flow fluctuations are present in the PS. These oscillations happen as a result of the slow response of the governor and nonlinearity of the system model. In such a situation, one of the most effective methods is the use of fast-active EES devices to solve this problem [24].

5.2.2 Modeling of the System Under Study

Two power systems (PSs) are considered in the present work. The first PS model is a three-area unequal non-reheat thermal traditional power system (System Model-1) [26] and the second PS is a two-area multi-source thermal-gas restructured/deregulated PS (System Model-2) [35]. The simulation models of both the introduced systems are described in Figure 5.7(a) and 5.7(b).

5.2.3 Control Approach

i. Controller
While designing a controller, the key characteristics that an LFC controller should acquire are set point following ability, robustness for any uncertainty, and curbing of load disturbances. The controller that is regularly used in many conventional or modern industries is a proportional-integral (PI) controller due to its simplicity, good performance, and cost-effectiveness. Therefore, PI controller is selected for LFC in this present work. Figure 5.8 presents the main building blocks of PI controller where K_p and K_I are the proportional and integral gains of the controller, respectively.

ii. Objective Function for Controller Design
The integral time absolute error (ITAE) is nominated as the objective function for controller design because it helps to tune a controller with better disturbance rejection abilities. The formulation of ITAE is given as follows:

$$J_{ITAE} = \int_0^{t_{sim}} \left[\sum_{i=1}^{N_{unit}} \left(|F_{Area-i}| + |P_{Tie-line,i}| \right) \right] \cdot t_{sim} dt \quad (5.12)$$

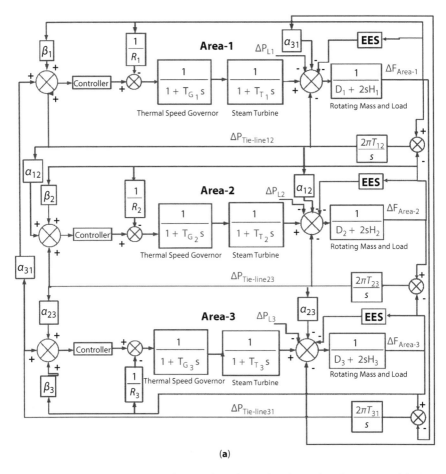

Figure 5.7 (a) Three-area unequal non-reheat thermal traditional PS (System Model-1).

where ΔF_{Area-i} and $\Delta P_{Tie-line,i}$ signifies the deviations in the frequency and the tie-line power in i^{th} area, respectively. t_{sim} is the time of simulation and N_{unit} is the count of areas interconnected in the system.

iii. Optimization Technique
Numerous optimization techniques are employed in LFC applications as reported in Table 5.1. However, the selection of the perfect algorithm is the toughest part for a given optimization problem as the no-free-lunch theorem specifies in [36]. Here, the moth flame optimization (MFO) technique proposed by Mirjalili [38] is selected as an optimization technique in this problem as it is already a well-established technique for LFC problems

Fast-Acting EES Systems for Frequency Regulation   123

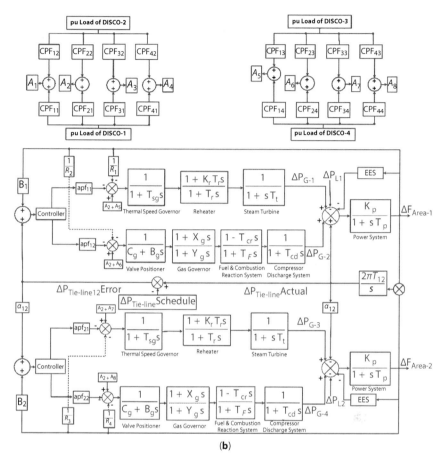

Figure 5.7 (b) Two-area multi-source thermal-gas deregulated PS (System Model-2).

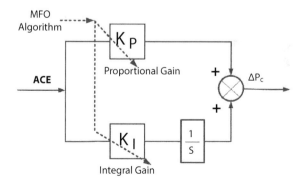

Figure 5.8 Structure of PI controller.

[4, 37]. In MFO, the lateral movement of moths around a bright object and their distinct investigation approaches at night are the key source of inspiration behind its working. Moths are insects, which use transverse orientation for navigation to fly at night. Both moths and flames play key roles in MFO algorithm and both of them are considered as a result of this algorithm. However, different methods are used to update the moths and the flames through each iteration. In MFO algorithm, Moths are the search agents in the considered search space and flames are the best position of moths in that search space. Thus, each moth flies in the search space, discovers a solution, and updates itself while coming across a superior solution. The mathematical analysis of MFO with more detail can be found in [38].

5.3 Impact of Fast-Acting EES Systems on the Frequency Regulation Services of Modern Power Systems

5.3.1 System Model-1

The traditional system (System Model-1) is analyzed with 1% (0.01 pu) load disruption in area-1. Simulations are executed for the system with and without the integration of EES (CES/SMES) devices. Figure 5.9(a-c) depicts the dynamic response of frequency and tie-line power. Further, time-domain parameters and performance indices for performed simulation are given in Table 5.2(a-c). The frequency oscillations reach at steady-state condition within 12.82 s for PI, 10.43 s for PI with SMES and 6.99 s PI for with CES for area-1, 11.03 s for PI, 9.47 s for PI with SMES and 6.72 s PI for with CES for area-2 and 13.03 s for PI, 9.78 s for PI with SMES and 8.06 s PI for with CES for area-3. Similarly, it is observed from Table 5.2(b) that tie-line power oscillations reach at steady-state condition within 9.48 s for PI, 7.96 s for PI with SMES and 6.72 s for PI with CES for area-1, 8.47 s for PI, 8.32 s for PI with SMES, and 6.72 s for PI with CES for area-2 and 11.03 s for PI, 9.38 s for PI with SMES and 6.86 s for PI with CES for area-3. Thus, it has been observed that the integration of CES mitigated the frequency and tie-line oscillations quickly compared to the system with/without SMES.

Comparative analysis of time performance index reveals that objective function values after integration of CES (IAE = 0.028761, ITAE = 0.17611, ISE = $9.667e^{-05}$ and ITSE = 0.00055852) are least compared to integration of SMES (IAE = 0.046630, ITAE = 0.33116, ISE = $13.156e^{-05}$ and ITSE = 0.00081055) and without EES (IAE = 0.076547, ITAE = 0.59104,

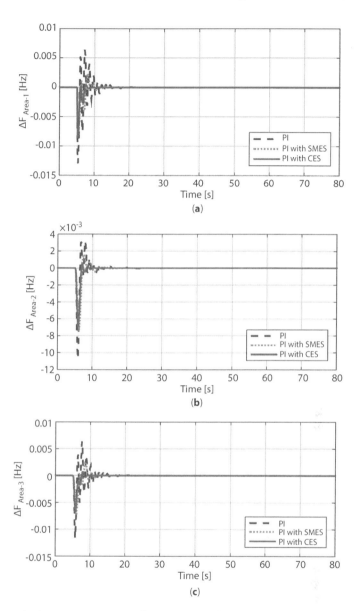

Figure 5.9 (a-c) Dynamic responses for system model-1. (*Continued*)

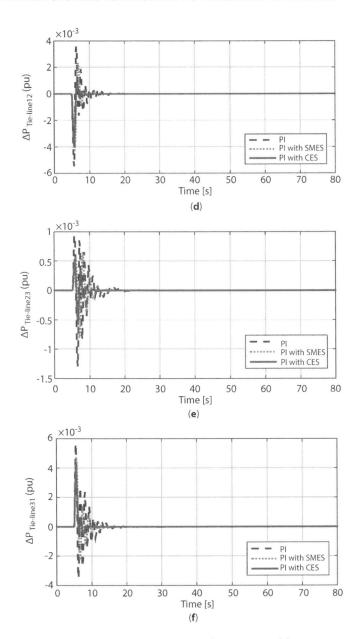

Figure 5.9 (Continued) (d-f) Dynamic responses for system model-1.

Table 5.2(a) Settling time, overshoot, and undershoot for frequency deviation for system model-1.

Controllers	Area 1			Area 2			Area 3		
	ST (s)	OS (Hz)	US (-ve) (Hz)	ST (s)	OS (Hz)	US (-ve) (Hz)	ST (s)	OS (Hz)	US (-ve) (Hz)
PI	12.82	0.0063	0.0129	11.03	0.00317	0.0106	13.03	0.0063	0.0115
PI with SMES	10.43	0.0022	0.0082	9.47	0.0015	0.0073	9.78	0.0024	0.0073
PI with CES	6.99	0.0004	0.0086	6.72	0.0004	0.0068	8.06	0.0007	0.0067

Table 5.2(b) Settling time, overshoot and undershoot for tie-line power deviation for system model-1.

Controller	$\Delta Ptie_{12}$			$\Delta Ptie_{23}$			$\Delta Ptie_{31}$		
	ST (s)	OS (pu)	US (-ve) (pu)	ST (s)	OS (pu)	US (-ve) (pu)	ST (s)	OS (pu)	US (-ve) (pu)
PI	9.48	0.0035	0.0054	8.47	0.0009	0.0012	11.03	0.0056	0.0035
PI with SMES	7.96	0.0023	0.0043	8.32	0.0006	0.0008	9.38	0.0047	0.0022
PI with CES	6.72	0.0009	0.0039	6.72	0.0005	0.0005	6.86	0.0042	0.0004

Table 5.2(c) Performance indices for system model-1.

Controller	IAE	ITAE	ISE	ITSE
PI	0.076547	0.59104	$28.705e^{-05}$	0.0018146
PI with SMES	0.046630	0.33116	$13.156e^{-05}$	0.00081055
PI with CES	0.028761	0.17611	$9.667e^{-05}$	0.00055852

ISE = $28.705e^{-05}$ and ITSE = 0.0018146). Thus, from the simulation outcome, it is evident that integration of CES yields finer enhancement in system performance with respect to settling time (ST), overshoot (OS), and undershoot (US).

5.3.2 System Model-2

The deregulated system (System Model-2) is analyzed with two diverse transactions, namely, bilateral and contract violation. ACE participation factors (apf) of every generation companies (GENCOs) is considered as, $apf_{11}=0.75$, $apf_{12}=0.25$, $apf_{21}=0.5$ and $apf_{22}=0.5$. Each distribution company (DISCO) has a load request of 0.005 pu.

i. Bilateral Transactions (BT)
In BT condition, all GENCOs and DISCOs don't have any restriction for settling a treaty among different areas [1]. Also, DISCOs have more chances to set up the agreement with all GENCOs of other areas and the same area. The contract of DISCOs with GENCOs is displayed via DISCO participation matrix (DPM) as follows:

$$DPM = \begin{bmatrix} 0.5 & 0.25 & 0 & 0.3 \\ 0.2 & 0.25 & 0 & 0 \\ 0 & 0.25 & 1 & 0.7 \\ 0.3 & 0.25 & 0 & 0 \end{bmatrix} \quad (5.13)$$

Total load request of all DISCOs should match with total generation from all GENCOs and generation of all GENCOs required according to the contract in (5.13) can be calculated as follows:

$\Delta P_{G-1} = 0.5 \times 0.005 + 0.25 \times 0.005 + 0 \times 0.005 + 0.3 \times 0.005 = 0.00525$ pu

$\Delta P_{G-2} = 0.2 \times 0.005 + 0.25 \times 0.005 + 0 \times 0.005 + 0 \times 0.005 = 0.00225$ pu

$\Delta P_{G-3} = 0 \times 0.005 + 0.25 \times 0.005 + 1 \times 0.005 + 0.7 \times 0.005 = 0.00975$ pu

$\Delta P_{G-4} = 0.3 \times 0.005 + 0.25 \times 0.005 + 0 \times 0.005 + 0 \times 0.005 = 0.00275$ pu

In considered DPM, off-diagonal elements demonstrated the contract that DISCOs made with GENCOs of other areas. The actual/scheduled steady-state tie-line power between areas is determined as in (5.14).

$$\Delta P_{\text{Tie-line}}^{\text{Actual}} = \sum_{i=1}^{2}\sum_{j=3}^{4} cpf_{ij}\Delta P_{Lj} - \sum_{i=3}^{4}\sum_{j=1}^{2} cpf_{ij}\Delta P_{Lj} \quad (5.14)$$

Here, DPM elements cpfs denote contract participation factors and ΔP_L is DISCO demand. Hence

$$\Delta P_{\text{Tie-line}}^{\text{Actual}} = (0+0)\times 0.005 + (0.3+0)\times 0.005 -$$
$$(0+0.3)\times 0.005 - (0.25+0.25)\times 0.005 = 0.0025 \text{ pu}$$

The tie-line power flow error is stated by (5.15):

$$\Delta P_{\text{Tie-line}}^{\text{Error}} = \Delta P_{\text{Tie-line}}^{\text{Actual}} - \Delta P_{\text{Tie-line}}^{\text{Scheduled}} \quad (5.15)$$

ii. Contract Violation (CV)
In CV transactions case, sometimes DISCOs demand surplus power against the agreement ethics. This additional uncontracted power must be delivered by GENCOs of the same area to whom the DISCOs belongs. Thus, surplus power is reflected only as an internal load of that local area. In this regard, the first DISCO of area-1 demands surplus power 0.003 pu at 40 s. Therefore, now the total load demand in area-1 is:

ΔP_{D1} = sum of contracted demand of $DISCO_1$ and $DISCO_2$ + uncontracted demand = 0.01 + 0.003 = 0.013 pu

To balance total load request of all DISCOs plus uncontracted load with the generation of all GENCOs, updated generation of all GENCOs can be calculated as below:

$\Delta P_{G-1} = 0.5 \times 0.005 + 0.25 \times 0.005 + 0 \times 0.005 + 0.3 \times 0.005 + 0.75 \times 0.003 = 0.0075$ pu

$\Delta P_{G-2} = 0.2 \times 0.005 + 0.25 \times 0.005 + 0 \times 0.005 + 0 \times 0.005 + 0.25 \times 0.003 = 0.003$ pu

$\Delta P_{G-3} = 0 \times 0.005 + 0.25 \times 0.005 + 1 \times 0.005 + 0.7 \times 0.005 + 0.5 \times 0 = 0.00975$ pu

$\Delta P_{G-4} = 0.3 \times 0.005 + 0.25 \times 0.005 + 0 \times 0.005 + 0 \times 0.005 + 0.5 \times 0 = 0.00275$ pu

The dynamic responses and comparative analysis of the system for different simulations have been displayed and recorded in Figure 5.10(a-d) and Table 5.3(a-b). The dynamic response of frequency for area-1 and area-2 reaches at steady state zero error within 15.27 s for PI, 10.55 s for PI with SMES, 9.65 s PI for with CES, and 14.82s for PI, 10.55 s for PI with SMES, 10.18 s for PI with CES, respectively. Further, undershoot oscillations of frequency for area-1 and area-2 are -0.0286 Hz for PI, -0.0128 Hz for PI with SMES, -0.0124 Hz for PI with CES, and -0.0311 Hz for PI, -0.0106 Hz for PI with SMES, -0.0102 Hz PI for with CES, respectively. In view of the simulation results, after integration of EES, system transient response shows finer outcome as contrasted to without EES. Further, the time domain parameters in Table 5.3(a) and dynamic responses in Figure 5.10(a-d) indicated that integration of CES enhances the system performance more compared to SMES and supports the dynamic stability of the system. Similarly, from Table 5.3(b), it is observed that the indices values after integration of CES (IAE = 0.038186, ITAE = 0.17497, ISE = 0.0001702 and ITSE = 0.0001443) are less compared to integration of SMES (IAE = 0.062375, ITAE = 0.23859, ISE = 0.00031236 and ITSE = 0.000428) and without EES (IAE = 0.18401, ITAE = 0.66764, ISE = 0.0027286 and ITSE = 0.0048073). Figure 5.11(a-b) shows the GENCOs output power profiles. The GENCOs generated power profile attained in Figure 5.11(a-b) is similar to the calculated values and relieved the variations in the system more rapidly. Thus, based on the simulation outcomes, the integration of CES gave a finer performance as compared to other proposed control strategies.

Figure 5.12(a-d) represents the frequency and tie-line power alternations alleviations without and with EES for CV transactions case. The comparative analysis of transient parameters and performance indices are indicated

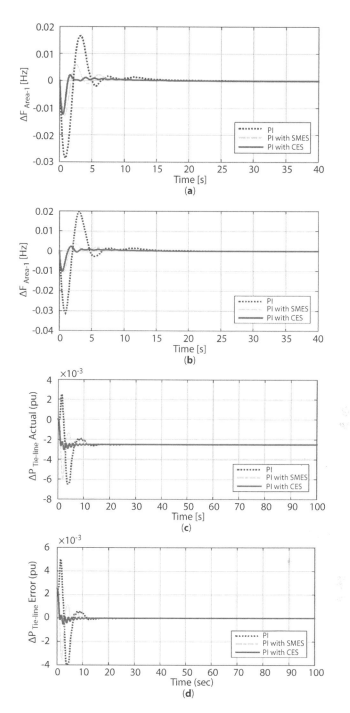

Figure 5.10 Dynamic responses under BT scenario.

Table 5.3(a) Performance indices for system model-2 under BT scenario.

Controller	Area 1			Area 2			$\Delta P_{Tie\text{-}line}^{Error}$		
	ST (s)	OS (Hz)	US (-ve) (Hz)	ST (s)	OS (Hz)	US (-ve) (Hz)	ST (S)	OS (pu)	UV (-ve) (pu)
PI	15.27	0.0165	0.0286	14.82	0.0197	0.0311	9.24	0.0050	0.0039
PI with SMES	10.55	0.0061	0.0128	10.55	0.0036	0.0106	4.74	0.0025	0.0035
PI with CES	9.65	0.0021	0.0124	10.18	0.0023	0.0102	2.40	0.0025	0.0005

Table 5.3(b) Performance indices for system model-2 under BT scenario.

Controller	IAE	ITAE	ISE	ITSE
PI	0.18401	0.66764	0.0027286	0.0048073
PI with SMES	0.062375	0.23859	0.00031236	0.000428
PI with CES	0.038186	0.17497	0.0001702	0.0001443

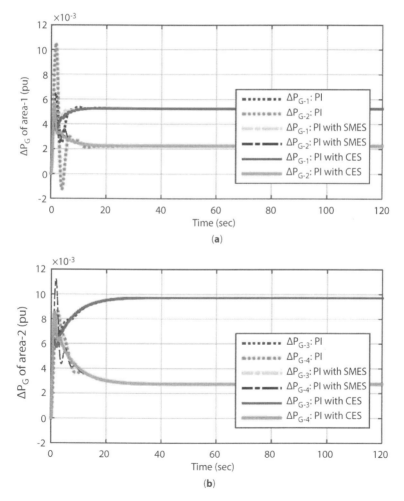

Figure 5.11 Dynamic responses of GENCOs output power under BT scenario.

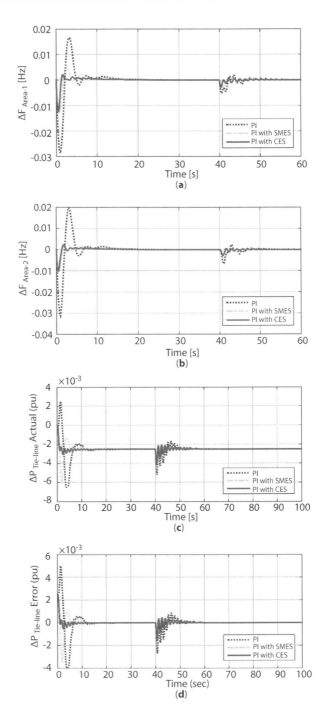

Figure 5.12 Dynamic responses under CV scenario.

Table 5.4(a) Performance indices for system model-2 under CV scenario.

Controller	Area 1				Area 2				$\Delta P_{Tie\text{-}line}^{Error}$	
	ST (s)	OS (Hz)	US (–ve) (Hz)		ST (s)	OS (Hz)	US (–ve) (Hz)		ST (S)	OS (pu)	UV (–ve) (pu)
PI	51.36	0.0165	0.0286		50.70	0.0197	0.0311		47.92	0.0050	0.0039
PI with SMES	43.38	0.0061	0.0128		43.16	0.0036	0.0106		43.06	0.0025	0.0035
PI with CES	44.33	0.0020	0.0124		43.68	0.0023	0.0102		42.11	0.0025	0.0015

Table 5.4(b) Performance indices for system model-2 under CV scenario.

Proposed controller	IAE	ITAE	ISE	ITSE
PI	0.21852	2.1926	0.0028097	0.0081933
PI with SMES	0.083026	1.1394	0.0003475	0.0018865
PI with CES	0.050478	0.71235	0.00018224	0.00064192

in Table 5.4(a-b). It has been observed that again in CV transactions, integration of EES reduces the settling time and supports the dynamic stability of the system. The profile of GNECOs generation is substantial in this transaction due to the existence of uncontracted surplus power demand. From the results, as represented in Figure 5.13 (a-b), GENCO$_1$ (0.0075 pu) and GENCO$_2$ (0.003 pu) meet the uncontracted load demand of DISCO$_1$. It is also pertinent to mention here that the uncontracted surplus power demand of DISCO$_1$ is not affected the generation of GENCO$_3$ (0.00975 pu) and GENCO$_4$ (0.00275 pu). Finally, the results disclose that the integration of SMES and CES in the power system improves the system's dynamic performance than without EES. Further, after the integration of CES, the system exhibits the finest dynamic performance and enhances the system's dynamic stability.

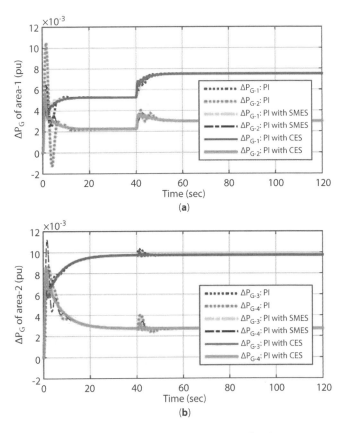

Figure 5.13 Dynamic responses of GENCOs output power under CV transactions scenario.

5.4 Conclusion

This chapter inspects the outcome of the integration of SMES and CES units in multi-area traditional and deregulated electric power systems for alleviating the frequency variation issues. The coordinated operation of the MFO tuned PI controller with CES/SMES is studied and tested in a three-area unequal thermal system and two-area thermal-gas deregulated system. The simulation results are assessed after an abrupt load disturbance in area-1 for test model-1 and both areas under bilateral transactions for test model-2. It is observed that in both cases, in the presence of SMES or CES, the system dynamic performance is upgraded in terms of settling time, peak overshoot, damping, etc., and the area frequency error reaches rapidly to zero. Hence, the SMES or CES can meritoriously diminish the frequency and tie-line power-related concerns in a traditional and deregulated electric power system. However, in the presence of CES, more improvement in system dynamics is observed compared to SMES in both cases.

Appendix A

Power system data

- System model-1 (Traditional unequal three-area non-reheat thermal power system) [26, 39]:
  $T_{G1} = 0.08s$, $T_{T1} = 0.4s$, $R_1 = 3$ Hz/pu, $\beta_1 = 0.3483$ pu/Hz, $2H_1 = 0.1667$ Hz/pu, $D_1 = 0.015$ pu/Hz, $T_{12} = 0.2$ pu/Hz, $a_{12} = -1$, $T_{G2} = 0.06s$, $T_{T2} = 0.44s$, $R_2 = 2.73$ Hz/pu, $\beta_2 = 0.3827$ pu/Hz, $2H_2 = 0.20177$ Hz/pu, $D_2 = 0.016$ pu/Hz, $T_{23} = 0.12$ pu/Hz, $a_{23} = -1$, $T_{G3} = 0.07s$, $T_{T3} = 0.30s$, $R_3 = 2.83$ Hz/pu, $\beta_3 = 0.3692$ pu/Hz, $2H_3 = 0.1247$ Hz/pu, $D_3 = 0.015$ pu/Hz, $T_{31} = 0.25$ pu/Hz, $a_{31} = -1$.
- System model-2 (Restructured two-area multisource thermal power system) [35]:
  $K_p = 120$ Hz/pu MW, $T_p = 20s$, $\beta = 0.4312$ pu MW/Hz, $R = 2.4$ Hz/pu MW, $a = 1$, $T_{12} = 0.02712$ pu MW/rad, $Y_g = 1.1$ s, $X_g = 0.6$ s, $a_{12} = -1$, $T_F = 0.239$ s, $T_{cd} = 0.2$ s, $B_g = 0.049$ s, $C_g = 1$, $T_{cr} = 0.01$ s, $K_r = 0.3$, $T_r = 10s$, $T_{sg} = 0.06$ s, $T_t = 0.3$ s.
  CES [35]:
  $K_{CES} = 0.3$, $T_{CES} = 0.0352$ s; $T_{C1} = 0.279$ s, $T_{C2} = 0.025$ s, $T_{C3} = 0.0411$ s, $T_{C4} = 0.1$ s.

SMES [40]:
K_{SMES}= 0.2035, T_{SMES}= 0.03, T_{S1}= 0.2333, T_{S2}= 0.016, T_{S3}=0.7087, T_{S4}=0.2481.

References

1. M. Sharma, S. Dhundhara, Y. Arya, and S. Prakash, "Frequency stabilization in deregulated energy system using coordinated operation of fuzzy controller and redox flow battery," *Int. J. Energy Res.*, Dec. 2020, doi: 10.1002/er.6328.
2. S. Dhundhara and Y. P. Verma, "Application of micro pump hydro energy storage for reliable operation of microgrid system," *IET Renew. Power Gener.*, vol. 14, no. 8, pp. 1368–1378, 2020, doi: 10.1049/iet-rpg.2019.0822.
3. S. Vazquez, S. Lukic, E. Galvan, L. G. Franquelo, J. M. Carrasco, and J. I. Leon, "Recent advances on Energy Storage Systems," in *IECON 2011 - 37th Annual Conference of the IEEE Industrial Electronics Society*, Nov. 2011, no. November, pp. 4636–4640, doi: 10.1109/IECON.2011.6120075.
4. S. Dhundhara and Y. P. Verma, "Grid frequency enhancement using coordinated action of wind unit with redox flow battery in a deregulated electricity market," *Int. Trans. Electr. Energy Syst.*, no. January 2019, pp. 1–33, 2020, doi: 10.1002/2050-7038.12189.
5. J. Abrell, S. Rausch, and C. Streitberger, "Buffering volatility: Storage investments and technology-specific renewable energy support," *Energy Econ.*, vol. 84, p. 104463, Oct. 2019, doi: 10.1016/j.eneco.2019.07.023.
6. O. Krishan and S. Suhag, "An updated review of energy storage systems: Classification and applications in distributed generation power systems incorporating renewable energy resources," *Int. J. Energy Res.*, no. July, pp. 1–40, 2018, doi: 10.1002/er.4285.
7. M. Jana, P. Sivakumar, M. Kota, M. G. Jung, and H. S. Park, "Phase- and interlayer spacing-controlled cobalt hydroxides for high performance asymmetric supercapacitor applications," *J. Power Sources*, vol. 422, pp. 9–17, May 2019, doi: 10.1016/j.jpowsour.2019.03.019.
8. M. Aneke and M. Wang, "Energy storage technologies and real life applications – A state of the art review," *Appl. Energy*, vol. 179, pp. 350–377, 2016, doi: 10.1016/j.apenergy.2016.06.097.
9. T. Kousksou, P. Bruel, A. Jamil, T. El Rhafiki, and Y. Zeraouli, "Energy storage: Applications and challenges," *Sol. Energy Mater. Sol. Cells*, vol. 120, no. PART A, pp. 59–80, 2014, doi: 10.1016/j.solmat.2013.08.015.
10. T. Mahto and V. Mukherjee, "A novel scaling factor based fuzzy logic controller for frequency control of an isolated hybrid power system," *Energy*, vol. 130, pp. 339–350, 2017, doi: 10.1016/j.energy.2017.04.155.
11. S. Dhundhara and Y. P. Verma, "Evaluation of CES and DFIG unit in AGC of realistic multisource deregulated power system," *Int. Trans. Electr. Energy Syst.*, vol. 27, no. 5, pp. 1–14, 2017, doi: 10.1002/etep.2304.

12. R. J. Abraham, D. Das, and A. Patra, "Automatic Generation Control of an Interconnected Power System with Capacitive Energy Storage," *World Acad. Sci. Eng. Technol.*, vol. 39, pp. 711–716, 2010.
13. R. J. Abraham, D. Das, and A. Patra, "Effect of Capacitive Energy Storage on Automatic Generation Control," *Power Eng. Conf. 2005. IPEC 2005. 7th Int.*, no. 1, pp. 1–5, 2003, doi: 10.1109/IPEC.2005.207066.
14. J. Raja and C. C. A. Rajan, "Stability Analysis and Effect of CES on ANN Based AGC for Frequency Excursion," *J. Electr. Eng. Technol.*, vol. 5, no. 4, pp. 552–560, 2010, doi: 10.5370/JEET.2010.5.4.552.
15. M. Ponnusamy, B. Banakara, S. S. Dash, and M. Veerasamy, "Design of integral controller for Load Frequency Control of Static Synchronous Series Compensator and Capacitive Energy Source based multi area system consisting of diverse sources of generation employing Imperialistic Competition Algorithm," *Int. J. Electr. Power Energy Syst.*, vol. 73, pp. 863–871, 2015, doi: 10.1016/j.ijepes.2015.06.019.
16. V. Knap, S. K. Chaudhary, D. I. Stroe, M. Swierczynski, B. I. Craciun, and R. Teodorescu, "Sizing of an energy storage system for grid inertial response and primary frequency reserve," *IEEE Trans. Power Syst.*, vol. 31, no. 5, pp. 3447–3456, 2016, doi: 10.1109/TPWRS.2015.2503565.
17. S. C. Tripathy, R. Balasubramanian, and P. S. C. Nair, "Small rating capacitive energy storage for dynamic performance improvement of automatic generation control," *Gener. Transm. Distrib. IEE Proc. C*, vol. 138, no. 1, pp. 103–111, 1991, doi: 10.1049/ip-c.1991.0013.
18. S. C. Tripathy and I. P. Mishra, "Dynamic performance of wind-diesel power system with capacitive energy storage," *Energy Convers. Manag.*, vol. 37, no. 12, pp. 1787–1798, 1996.
19. D. O. Akinyele and R. K. Rayudu, "Review of energy storage technologies for sustainable power networks," *Sustain. Energy Technol. Assessments*, vol. 8, pp. 74–91, 2014, doi: 10.1016/j.seta.2014.07.004.
20. A. Pappachen and A. Peer Fathima, "Load frequency control in deregulated power system integrated with SMES-TCPS combination using ANFIS controller," *Int. J. Electr. Power Energy Syst.*, vol. 82, pp. 519–534, 2016, doi: 10.1016/j.ijepes.2016.04.032.
21. P. Bhatt, S. P. Ghoshal, and R. Roy, "Coordinated control of TCPS and SMES for frequency regulation of interconnected restructured power systems with dynamic participation from DFIG based wind farm," *Renew. Energy*, vol. 40, no. 1, pp. 40–50, 2012, doi: 10.1016/j.renene.2011.08.035.
22. E. Schempp and W. D. Jackson, "Systems considerations in capacitive energy storage," *IECEC 96. Proc. 31st Intersoc. Energy Convers. Eng. Conf. Washington, DC, USA*, vol. 2, pp. 666–671, 1996, doi: 10.1109/iecec.1996.553777.
23. R. J. Abraham, D. Das, and A. Patra, "Automatic generation control of an interconnected power system with capacitive energy storage," *World Acad. Sci. Eng. Technol.*, vol. 30, pp. 711–716, 2010, doi: 10.1109/PQ.1998.710361.

24. S. M. Nosratabadi, M. Bornapour, and M. A. Gharaei, "Grasshopper optimization algorithm for optimal load frequency control considering Predictive Functional Modified PID controller in restructured multi-resource multi-area power system with Redox Flow Battery units," *Control Eng. Pract.*, vol. 89, no. June, pp. 204–227, Aug. 2019, doi: 10.1016/j.conengprac.2019.06.002.
25. A. Prakash, S. Murali, R. Shankar, and R. Bhushan, "HVDC tie-link modeling for restructured AGC using a novel fractional order cascade controller," *Electr. Power Syst. Res.*, vol. 170, pp. 244–258, May 2019, doi: 10.1016/j.epsr.2019.01.021.
26. M. Sharma, R. K. Bansal, and S. Prakash, "Robustness Analysis of LFC for Multi Area Power System integrated with SMES–TCPS by Artificial Intelligent Technique," *J. Electr. Eng. Technol.*, vol. 14, no. 1, pp. 97–110, Jan. 2019, doi: 10.1007/s42835-018-00035-3.
27. A. Gomaa Haroun and L. Yin-Ya, "A novel optimized fractional-order hybrid fuzzy intelligent PID controller for interconnected realistic power systems," *Trans. Inst. Meas. Control*, vol. 41, no. 11, pp. 3065–3080, Jul. 2019, doi: 10.1177/0142331218820913.
28. S. Dhundhara and Y. P. Verma, "Capacitive energy storage with optimized controller for frequency regulation in realistic multisource deregulated power system," *Energy*, vol. 147, pp. 1108–1128, 2018, doi: 10.1016/j.energy.2018.01.076.
29. M. Sharma, R. K. Bansal, S. Prakash, and S. Asefi, "MVO Algorithm Based LFC Design of a Six-Area Hybrid Diverse Power System Integrating IPFC and RFB," *IETE J. Res.*, vol. 0, no. 0, pp. 1–14, Dec. 2018, doi: 10.1080/03772063.2018.1548908.
30. W. Tasnin and L. C. Saikia, "Comparative performance of different energy storage devices in AGC of multi-source system including geothermal power plant," *J. Renew. Sustain. Energy*, vol. 10, no. 2, 2018, doi: 10.1063/1.5016596.
31. J. Morsali, K. Zare, and M. Tarafdar Hagh, "A novel dynamic model and control approach for SSSC to contribute effectively in AGC of a deregulated power system," *Int. J. Electr. Power Energy Syst.*, vol. 95, pp. 239–253, 2018, doi: 10.1016/j.ijepes.2017.08.033.
32. R. Rajbongshi and L. C. Saikia, "Performance of coordinated FACTS and energy storage devices in combined multiarea ALFC and AVR system," *J. Renew. Sustain. Energy*, vol. 9, no. 6, pp. 064101–21, Nov. 2017, doi: 10.1063/1.4986889.
33. K. R. Sudha and R. Vijaya Santhi, "Load Frequency Control of an Interconnected Reheat Thermal system using Type-2 fuzzy system including SMES units," *Int. J. Electr. Power Energy Syst.*, vol. 43, no. 1, pp. 1383–1392, Dec. 2012, doi: 10.1016/j.ijepes.2012.06.065.
34. A. Pappachen and A. P. Fathima, "Impact of SMES–TCSC Combination in a Multi-Area Deregulated Power System with GA-Based PI Controller," *J. Control. Autom. Electr. Syst.*, vol. 30, no. 6, pp. 1069–1081, Dec. 2019, doi: 10.1007/s40313-019-00492-9.

35. Y. Arya, "Effect of energy storage systems on automatic generation control of interconnected traditional and restructured energy systems," *Int. J. Energy Res.*, vol. 43, no. 12, pp. 6475–6493, Oct. 2019, doi: 10.1002/er.4493.
36. D. H. Wolpert and W. G. Macready, "No free lunch theorems for optimization," *IEEE Trans. Evol. Comput.*, vol. 1, no. 1, pp. 67–82, Apr. 1997, doi: 10.1109/4235.585893.
37. M. Sharma, R. K. Bansal, S. Prakash, and S. Dhundhara, "Frequency regulation in PV integrated Power System using MFO tuned PIDF controller," *2018 IEEE 8th Power India Int. Conf.*, pp. 1–6, 2019, doi: 10.1109/POWERI.2018.8704453.
38. S. Mirjalili, "Moth-flame optimization algorithm : A novel nature-inspired heuristic paradigm," *Knowledge-Based Syst.*, vol. 89, pp. 228–249, 2015, doi: 10.1016/j.knosys.2015.07.006.
39. H. Bevrani, *Robust Power System Frequency Control*. Springer, 2009.
40. A. Kumar and S. Suhag, "Effect of TCPS, SMES, and DFIG on load frequency control of a multi-area multi-source power system using multi-verse optimized fuzzy-PID controller with derivative filter," *JVC/Journal Vib. Control*, vol. 24, no. 24, pp. 5922–5937, 2018, doi: 10.1177/1077546317724968.

6
Solid-Oxide Fuel Cell and Its Control

Preeti Gupta*, Vivek Pahwa and Yajvender Pal Verma

Department of Electrical and Electronics Engineering, University Institute of Engineering and Technology, Panjab University, Chandigarh, India

---

Abstract

For reliable operation, battery energy storage system (BESS) with renewable energy source (RES), such as a wind and solar-based distribution system generally gives satisfactory performance when connected to AC grid. However, large charging time and requirement of one additional unit for recharging of batteries, and the intermittent nature of RESs are the main challenges associated with this system. These problems can be solved with non-intermittent fuel cells that do not require recharging. There are various kinds of fuel cells and among them the solid oxide fuel cell (SOFC) is the most efficient on the distribution system. However, SOFC is a multiple-input-multiple-output and non-linear device. There exists a strong coupling between its control variables, and when it is connected to the AC grid, the control problem becomes severe, due to the addition of variables associated with inverter and its control circuitry. Therefore, firstly, the characteristics of SOFC have been studied, by linearising it in the MATLAB/SIMULINK environment. Finally, after selecting the appropriate control variable pairing, a systematic development of a coordinated control scheme has been accomplished for improving the life of SOFC by keeping its associated variables into the feasible operating limits. Comparison of simulation results shows the superior performance of proposed scheme in contrast to the already existing control scheme. With this proposed scheme, it is possible to keep the grid voltage and grid frequency at one per unit. Further, this chapter also highlights the worldwide recent trend of fuel cells with their techno-economic aspects and market policies.

Corresponding author: er.guptapreeti07@gmail.com

Keywords: Fuel cells, solid-oxide fuel cell, dynamic performance of SOFC, co-ordination control, recent trends in fuel cells

Abbreviations

RES	Renewable Energy Source
SOFC	Solid-oxide fuel cell
MIMO	Multiple-input Multiple-output
ICEs	Internal combusting engines
AC	Alternate Current
DC	Direct Current
AFC	alkaline fuel cell
PEMFC	polymer exchange membrane fuel cell
DMFC	direct methanol fuel cells
PAFC	phosphoric acid fuel cells
MCFC	molten carbonate fuel cells
A: and C:	anode and cathode
FOA	feasible operating area
PWM	pulse width modulation
PLL	phase locked loop
R&D	research and development
PI	proportional and integral
GPC	generalized predictive control
NNPC	neural network predictive control
PSO	particle swarm optimization
ANFIS	Artificial Neural-Network Fuzzy Inference System
BESS	Battery energy storage system

Symbols and Molecular Formulae

e^-	-	electron
G	-	gaseous
li	-	lithium
H	H_2	hydrogen
O	O_2	oxygen
-	CO_2	Carbon-dioxide
-	CO	Carbon-mono-oxide
-	H_2O	water
-	CH_4	Methane or natural gas
-	CH_3OH	methanol

Nomenclature

li-ion	Lithium-ion
V_{sofc}	output SOFC voltage
V_{sofc}^{f}	standard potential of one cell in SOFC
V_{sofc}^{f}	generated potential of SOFC
R	gas-constant
T	operating temperature
F	faraday constant
p_{H_2}, p_{O_2} and p_{H_2O}	effective partial pressures of H_2, O_2 and H_2O in the cells
r	resistance
V_{drop}	ohmic drop
I_{sofc}	output SOFC current
$f_{H_2}^{in}$ and $f_{O_2}^{in}$	fuel flows of H_2 and O_2 at the inlet of SOFC
M_H, M_O and M_{H_O}	are molar constants of H_2, O_2 and H_2O
τ_H, τ_O and τ_{H_O}	response time constants of H_2, O_2 and H_2O
k_r	reaction constant
$f_{H_2}^{unc}$ and $f_{H_2}^{c}$	generated and controlled fuel feed of H_2
$f_{H_2}^{r}$	natural fuel flow of H_2
U_f and U_f^{opt}	generated and optimum fuel utilization factor
P_{sofc}	output SOFC power
aer	air excess ratio
r_{H_O}	stoichiometric ratio of H:O
τ_f	response time of fuel at inlet
K_p and K_i	proportional and integral constants
P_{ref} and P_g	reference and grid power
V_{abc} and I_{abc}	grid side voltage and current
f and ω	grid side frequency and angular frequency

6.1 Introduction

In the present scenario, power generation from renewable sources such as wind, solar and fuel cell has gained momentum owing to fast depletion of fossil fuels, increasing pollution level and the pressing need for a sustainable energy solution to meet the ever-increasing demand for electricity. Out of these sources of energy, solar and wind are freely available, but these are highly intermittent by nature [1]. Efforts have been made to remove this intermittency problem by adding an energy storage device, i.e., battery [2]. However, a battery suffers from the problems of large charging time and requirement of one extra source for charging, increasing its complexity and cost. This combination of wind/solar-based power plant with BESS can be replaced with fuel cell because this single device is renewable, requires no additional infrastructure for charging and has the ability to store [3]. It converts the chemical energy into surplus electrical energy to fulfill the demand of power supply. Additionally, it has no rotating part, resulting in less maintenance and nearly zero noise in comparison to the wind energy–based power plants, particularly [4, 5].

Apart from the above-mentioned advantages, fuel cells are sustainable, fuel flexible, low carbon emissive, and inexpensive. As a result they are gradually coming up as an excellent alternative in power systems as well as in other domestic and industrial applications [6, 7]. Figure 6.1 reveals that during the last decade, there has been an appreciable increase in the usage of this device, worldwide. It can be seen that the usage is more in Asia in comparison to rest of the world [8, 9].

In Asia, India has many ongoing and upcoming projects like "high pressure hydrogen cylinder" developed by IIT Kharagpur, which increases the energy storage density over existing cylinders [10]. Hence, owing to these merits, and wide range of applications, fuel cells have a significant role to play in the future. Many types of fuel cells such as alkaline fuel cell (AFC), polymer exchange membrane fuel cell (PEMFC), and solid oxide fuel cell (SOFC) [6, 7] are available in the market.

Among all the fuel cells, SOFCs are efficiently reliable on distribution networks [11–13]. These are natural gas-fed fuel cells that convert

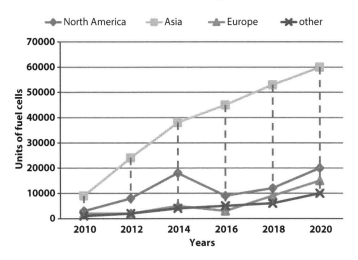

Figure 6.1 The use of fuel cells in various parts of the world.

hydrocarbons such as methane to DC electrical power. It is the most recent fuel cell that has been growing over the past two decades. So far, it has been discussed regarding its materialistic stresses based on the electrochemical and thermo-dynamical processes. Further, it is being exploited on designing its structure [14], material stress [15], physical modeling [16, 17], and various control-based modelings [18, 19]. In controlling, the variables like SOFC voltage, fuel input, partial pressures of oxygen-hydrogen, air excess ratio, utilization factor, temperature, etc., of SOFC are exploited [20–26]. Association of a number of variables makes it a complex device from analysis as well as from the control point of view. Therefore, there is a requirement to analyze the characteristics of SOFC and select its proper control variable pairing.

Further, the control becomes more complicated when this complex fuel cell is connected with the AC utility grid, because, this is producing DC power and it can be connected to AC grid through inverter. Again, this inverter and its associated control circuitry has a higher number of different control variables [27]. Therefore, there is a strong need to develop a coordinated control scheme between the SOFC and inverter, having a number of variables associated with each device, while connecting to the AC utility grid.

Therefore, in this chapter, the analysis of SOFC has been carried out to study its complex dynamic characteristics. And then a step-by-step development of a coordinated control scheme has been done for improving the life of this fuel cell by keeping its associated variables within feasible operating limits. Additionally, this chapter also introduces the fuel cell technologies and their different types in the market; their advantages and disadvantages along with their myriads of applications are presented. A brief discussion is also given on the recent trends in fuel cells, their techno-economic comparison, and the problems associated in deploying them in the market or to the modern power system.

6.2 Fuel Cells

A fuel cell is a multidisciplinary technology that involves chemical, thermal, and electrical aspects. It is an electrochemical device, unlike batteries in that it does not require recharging. Without diving into the history and origin of fuel cells, such as details of its evolution and structure formation, it can be simply given as generation of an electrical energy from the electrochemical processes as shown in Figure 6.2 [4, 7]. It represents the basic structure of a fuel cell that has two electrodes, separated by the electrolyte and connected in an external circuit. The electrodes are exposed to oxidant

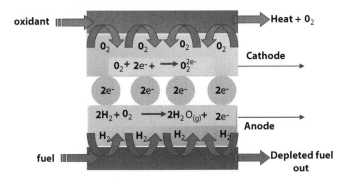

Figure 6.2 A basic structure of a fuel cell.

and hydrogen extracted from the fuel to produce the flow of electrons (e^-) and generate DC power at the cathode-anode terminals [28].

In Figure 6.2, the general reactions at cathode-anode electrodes are also represented. At anode, the hydrogen, (H_2) combine with oxygen, (O_2) to generate electrons (e^-). These ions travel towards the anode to participate with O_2 to form a negative molecule, (O_2^{2e-}). During this process, the heat is librated and the flow of directional is captured as the DC current.

6.2.1 Different Types of Fuel Cells

Generally, the type of electrolyte being used in a fuel cell defines its type or kind, such as SOFC which uses a ceramic material as the electrolyte [29]. Different types of fuel cells along with their electrolyte, anode-cathode reactions and the operating temperature are given in Figure 6.3 [4, 28].

Among all these fuel cells, high temperature fuel cells, like SOFCs or MCFCs, are efficient, with less impact on the environment [30]. PEMFCs and SOFCs are preferred due to high flexibility and low emissions. Thus, it is greatly contributing to society by delivering power and helping to reduce global warming; a lot of applications can be seen, as will be discussed further. Further, SOFCs are most reliable and efficient on distribution network [7, 12, 13, 31].

6.2.2 Advantages and Disadvantages

Fuel cells generate power through a mechanism that does not require combustion, unlike oil-based power plants. Since the concern about lowering environmental pollution is driving the need for improving power production, it is vital to note the following advantages of its energy's production.

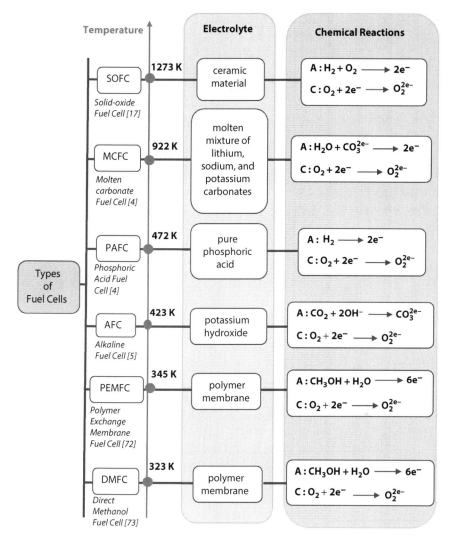

Figure 6.3 Types of fuel cells.

- Fuel cells are a highly efficient and reliable source of power. They provide efficiency up to 60-70% and higher in certain applications.
- They have the facility of fuel flexibility, i.e., biogas, natural gas, solar, hydrogen, etc,, are used as fuel-to-fuel cells.
- It requires low maintenance in comparison to the oil-based power systems because it does not require oil change or constant looking after.

- It gives a quiet operation because it has no rotational part; as a result, a noise-free operation is achieved.
- It gives reduced emissions during power production processes; thus the air quality is improved in comparison to other thermal power plants [32].

Although fuel cells are quite reliable, during co-generations they provide efficiency up to 35% only [32], which is low. Further, its future growth depends on the improvements in its fuel energy infrastructure and the rate of decline in utilizing the other polluting energies such as oil-based power stations. Since the concern for lowering environmental pollution is increasing the need for improving the infrastructure of energy production, distribution, and safe regulation, it is vital to emphasis the synthesis of fuels such as hydrogen, alcohol, and natural gases so that the fuel is directly fed to the fuel processor [33]. The practical generation and storage of these fuels is cumbersome because they oxidize and contain poor power densities [34].

6.2.3 Applications in Modern Power System

Despite the above-mentioned disadvantages, the emission-free automobiles and power plants will be more profitable and beneficial, as a number of applications exist for these fuel cell technologies as given in Table 6.1. In this, the different fuel cells that have the respective applications are mentioned along with a few examples.

The main application in the field of power system concerns electricity production and co-generation for industrial and commercial purposes, mainly due to the increase in environmental concerns, ever-increasing electricity consumption, and reforms in market policies. It has led to the diversification in the sources of power supply and has raised a substantial requirement for the production of quality power. Presently, these fuel cells are used as a source for continuous power backup/uninterrupted power supply or as a low power plant that utilizes PEMFC. However, fuel cells have a lot more potential than what has been explored so far, because highly efficient fuel cells like SOFC have still more scope and are the future of the modern power system.

6.3 Solid-Oxide Fuel Cell

As per the above literature, the most attractive option for a modern power system is fuel cell as it finds a wide range of domestic and industrial applications due to the inherent properties such as renewable, highly efficient,

Table 6.1 Applications and examples of different fuel cells.

S. no.	Fuel cells	Application	Examples
1.	AFC and MFC	Space stations	Daily station life such as cooking, drinking water, etc.
2.	PEMFC	Transportation power generations	Powered bicycles, buses, recreation vehicles, etc. [35, 36]
3.	SOFC, PAFC and MCFC	Stationary power stations	Emission-free power stations
4.	DMFC	Portable power generations	Replacement of conventional batteries such as li-ion battery, etc. [37, 38]

lower carbon emissions, and inexpensiveness. Among fuel cells, SOFCs are more reliable and efficient on distribution networks [11–13]. Its prominent application as a stationary power plant allows internal reforming of the fuel in the fuel processor that benefits the system economically [39]. It undergoes chemical reactions which includes both thermal and electrochemical processes [31, 40, 41].

The fuel input to SOFC is any hydrocarbon compound such as methane from which H_2 is extracted at anode as shown in Figure 6.4. It can be seen that the basic structure of SOFC represents the set of reactions occurring at the cathode-anode terminals. At the inlet two fuel flow inputs exist, i.e., the fuel flow of hydrogen, $f_{H_2}^{in}$ and oxygen, $f_{O_2}^{in}$. The rate at which the fuel is utilized in the SOFC is defined as its fuel utilization factor, U_f and the rate at which the oxygen is fed to the SOFC is its air excess ratio, aer [42–44]. At the outlet very less amount of carbon-di-oxide, CO_2 and heat is released along with the output voltage, V_{sofc} at the cathode-anode terminal.

Various researchers have modeled this SOFC structure for its performance analyzes in their respective domains [45–47]. The modeling of a SOFC using both lumped and distributed modeling approaches for a real-time emulation and control is discussed [41]. A quasi-two-dimensional physically based model has been proposed [40] for the description of transport and reaction in planar SOFC.

Apart from the electrochemistry and transport phenomena in the cells for the various constraints like temperature, a SIMULINK-based model for analyzing the electrical dynamic behavior of this fuel cell has been suggested [16, 48].

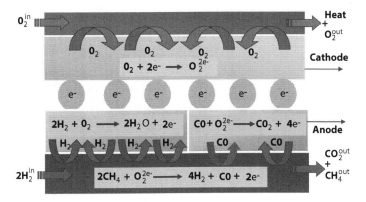

Figure 6.4 A basic structure of solid-oxide fuel cell.

6.3.1 Mathematical Modeling

An appropriately designed dynamical model of SOFC for investigating its performance in electrical as well as feasible operation, its mathematical model has been taken into consideration as shown in Figure 6.5 [16].

During the reaction, V_{sofc} across the cathode-anode terminals of SOFC is expressed using a Nernst equation during open circuit condition as mentioned in (6.1),

$$V_{sofc}\big|_{open} = V_{sofc}^o + Nernst \qquad (6.1)$$

where, $Nernst = V_{sofc}^f \ln\left[\dfrac{(p_{H_2})^2 p_{O_2}}{(p_{H_2O})^2}\right]$, $V_{sofc}^f = \dfrac{RT}{4F}$ and $p_{H_2} = \dfrac{1}{M_H}\left[f_{H_2}^{in} - \dfrac{I_{sofc}}{2F}\right]$,

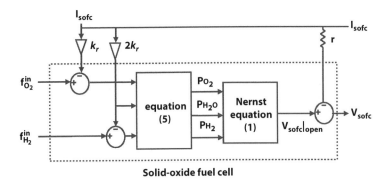

Figure 6.5 The dynamic model of SOFC.

$$p_{O_2} = \frac{1}{M_O}\left[f_{O_2}^{in} - \frac{I_{sofc}}{4F}\right], \text{ and } p_{H_2O} = \frac{1}{M_{H\_O}}\left[f_{H_2O}^{in} - \frac{I_{sofc}}{2F}\right] \quad [16, 19, 22].$$

Similarly, for n number of cells in SOFC, the open circuit voltage will be n $V_{sofc}|_{open}$ for the reaction constant, $k_r = \frac{0.25n}{F}$ [16].

6.3.2 Linearization

During the load change, V_{sofc} reduces due to three types of potential losses occurring in the cells [19], namely, activation loss, ohm loss, and concentration loss. These losses are represented by the loss across a resistance, r thus equivalent change in ohmic drop, ΔV_{drop} is represented in (6.2).

$$\Delta V_{drop} = r\Delta I_{sofc} \quad (6.2)$$

where, ΔI_{sofc} is the generated change in direct current of SOFC. Therefore, the final change in output voltage of SOFC is ΔV_{sofc} as given in (6.3).

$$\Delta V_{sofc} = \Delta V_{sofc}|_{open} - \Delta V_{drop} \quad (6.3)$$

Substituting (6.1) and (6.4) in (6.3),

$$\Delta V_{sofc} = V_{sofc}^{f}\,\Delta \ln\left[\frac{(p_{H_2})^2\, p_{O_2}}{(p_{H_2O})^2}\right] - r\Delta I_{sofc} \quad (6.4)$$

where, $\Delta V_{sofc}^{o} = 0$ because the SOFC is assumed to be operating at the constant operating temperature where it performs for the highest efficiency and p_{H_2}, p_{O_2} and p_{H_2O} of H_2, O_2 and H_2O in the cells further depend on the change in fuel flow of hydrogen, $\Delta f_{H_2}^{in}$ and oxygen, $\Delta f_{O_2}^{in}$ at the inlet of cell and ΔI_{sofc} as expressed in (6.5),

$$\begin{bmatrix}\Delta p_{H_2}(s)\\ \Delta p_{O_2}(s)\\ \Delta p_{H_2O}(s)\end{bmatrix} = \begin{bmatrix}\frac{1}{M_H(1+s\tau_H)}\\ 0\\ 0\end{bmatrix}\Delta f_{H_2}^{in}(s) + \begin{bmatrix}0\\ \frac{1}{M_O(1+s\tau_O)}\\ 0\end{bmatrix}\Delta f_{O_2}^{in}(s)$$

$$+ \frac{1}{2F}\begin{bmatrix}-1\\ -1\\ \frac{1}{M_{H\_O}(1+s\tau_{H\_O})}\end{bmatrix}\Delta I_{sofc}(s) \quad (6.5)$$

Now the dynamic model of SOFC represented in Figure 6.5 is linearized using (6.4). It is implemented in MATLAB/SIMULINK environment with the data/values of all the model parameters given in Table 6.2 [20]. With this, the Bode plot of SOFC for I_{sofc} = 0.1 p.u. and 1.0 p.u. have been obtained in Figure 6.6. The input is a small change in fuel feed of hydrogen, $f_{H_2}^{unc}$ whereas the output is the change in SOFC voltage, ΔV_{sofc}. It can be observed from the plot that by changing the current, the change of crossover frequency is negligible. The zoom view of magnitude plot of bode has been drawn for more clarity. Therefore, it is a clear observation that the SOFC is a slow responding device.

Another vitality for the SOFC is to operate within a feasible area for its secured life [23]. The FOA specifies boundaries within which it must operate for its safe operation are based on its fuel utilization factor, U_f, output voltage, V_{sofc}, output current, I_{sofc} and active power of SOFC, P_{sofc}. Because during any

Table 6.2 Values of the parameters used in SOFC.

SOFC parameters	Values
T (K)	1273
F (C/kmol)	96.487e + 06
R (J/kmol-K)	8.314
V_{sofc}^{o} (volts)	1.18
k_r (kmol/s-A)	1.166e - 06
r (ohms/cell)	3.2813e - 04
M_H (kmol/s-atm)	8.343e - 4
M_{H_O} (kmol/s-atm)	2.81e - 4
M_O (kmol/s-atm)	2.52e - 3
τ_H (s)	26.1
τ_{H_O} (s)	78.3
τ_O (s)	2.91
τ_f (s)	5
r_{H_O}	1.145
U_f^{opt}	0.8

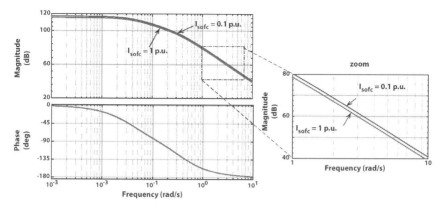

Figure 6.6 The bode plot.

operation, the SOFC should never be over or under used, therefore its fuel utilization must be within the limits. The output voltage must also remain within the prescribed limits as per the AC load requirements, and the SOFC must operate within its rated output power [22]. As a result, the FOA boundaries limits are specified for the SOFC for a successful operation as given in (6.6).

$$0.7 \leq U_f \leq 0.9; 0.95 \leq V_{sofc} \leq 1.05; 0.1 \leq I_{sofc} \leq 1; 0.1 \leq P_{sofc} \leq 1; \tag{6.6}$$

6.3.3 Control Schemes for Solid-Oxide Fuel Cell Based Power System

This MIMO storage device, SOFC has number of variables, i.e., U_f, $f_{H_2}^{in}$, $f_{O_2}^{in}$, I_{sofc}, aer, T etc., that makes it a non-linear device. In addition, there exists a strong coupling between each variable [15]. The main challenge for its control is the selection of appropriate variables. Recent researches have shown satisfactory performance of this device at a temperature of 1273K [42, 43] and keeping H to O ratio as r_{H_2O} of $f_{H_2}^{in}$ [43, 49]. There is explicit relation between $f_{H_2}^{in}$, I_{sofc} and U_f as given by (6.7),

$$U_f = \frac{2 k_r I_{sofc}}{f_{H_2}^{in}} \tag{6.7}$$

and between $f_{O_2}^{in}$, I_{sofc} and *aer* as given by (6.8).

$$aer = \frac{f_{O_2}^{in}}{k_r I_{sofc}} \tag{6.8}$$

After selecting the control variables, the literature [20, 22] dictates that it is not possible to control both the output voltage and fuel utilization factor simultaneously. Therefore, the control schemes for the SOFC are subdivided in two types as follows:

- Constant voltage control
- Constant fuel utilization control

6.3.3.1 Constant Voltage Control

In a power system, it is necessary to operate for constant voltage irrespective of the type of power source used in the system. When the system is subjected from the no load to load, it experiences a rise in loss as well due to which the drop in voltage is a natural phenomenon. Here, the source is DC-type that is not sustainable until connected to the AC utility grid. Nowadays it is taken into consideration through the inverters [50, 51]. These are the main interfacing systems between the renewable source of power and the network as they help in synchronizing the system. Also, they convert the DC power produced by the renewable energies to AC power [52, 53] and fulfill the needs such as voltage and frequency stability necessary for a grid connection [54, 55]. It is vital that inverter satisfy the need because if it is not capable of handling the constant voltage, then matching the rate of power generation with the traditional power system's frequency becomes tedious. However, the constant voltage control is easily attainable by any PWM technique in inverters [20]. Therefore, it is not discussed as the focus here is the control of grid-connected SOFC.

6.3.3.2 Constant Fuel Utilization Control

In the recent past, this control scheme has been used as feed-forward control law because it is easy to implement and the cost involved is also less [22, 23]. For the sake of completeness, this scheme has been developed here and Figure 6.7 shows its implementation.

It can be seen that $f_{H_2}^{in}$ at the inlet of SOFC differs from the fuel feed because there exist a fuel processor. It is a first order transfer function with the response time constant of fuel [22]. This processes the fuel and transfers the processed fuel by positioning it in the right direction so that the fuel participates properly in the chemical reactions [40]. Therefore, $f_{H_2}^{unc}$ is calculated for U_f^{opt} as given in (6.9),

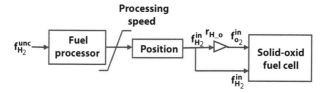

Figure 6.7 A schematic for a control scheme of 'constant utilization factor' for SOFC-based power system.

$$f_{H_2}^{unc} = \frac{2k_r I_{sofc}}{U_f^{opt}} \qquad (6.9)$$

and, the fuel at the inlet is the processed fuel, $f_{H_2}^{in}$ given in (10).

$$f_{H_2}^{in} = \frac{f_{H_2}^{unc}}{1 + s\tau_f} \qquad (6.10)$$

Thereafter, the oxidant is supplied as per the required H to O ratio, which completes the electro-chemical process of energy conversion [16]. The fuel flow of oxygen, $f_{O_2}^{in}$ at the inlet of SOFC is given in (6.11).

$$f_{O_2}^{in} = r_{H\_O} f_{H_2}^{in} \qquad (6.11)$$

This scheme is applied to a grid-connected SOFC system for observing the responses of SOFC. This system is developed and implemented in MATLAB/SIMLINK environment as shown in Figure 6.8.

It can be seen that SOFC with its fuel inputs is connected to the grid via DC-link capacitor, C and inverter. This capacitor, C provides low impedance path for the turned-off current, acts as an energy storage device and balances the power difference between SOFC and inverter [56]. And the inverter converts DC to AC power to provide sustainability to SOFC

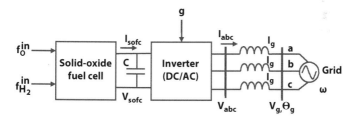

Figure 6.8 A grid-connected SOFC system.

through a grid connection. The additional components used in this system are reference power loop, and power conditioning unit, which are essential for connecting this SOFC to AC utility grid [20]. The brief functioning of these additional components are given as:

- *Reference power loop:* The function of this unit is to generate a reference current signal for the operation of inverter.
- *Power conditioning unit:* This unit is composed of inverter, inverter control strategy, PLL and, voltage and current measurement units.

This system is subjected to the load change of 100% and 50% as shown in Figure 6.9 where a step change in input reference power from 0.1 p.u. to 1 p.u. at 1 second and from 1 p.u. to 0.5 p.u. at 12 second. And the total simulation time taken into consideration is 25 seconds.

The responses of SOFC including the fuel flows of hydrogen, $f_{H_2}^{in}$ and oxygen, $f_{O_2}^{in}$ at the inlet, their utilization factor, U_f and excess ratio, aer are shown in Figure 6.10. In Figure 6.10(a), it is seen that the increase in $f_{H_2}^{in}$ is very slow due to slow pole associated with the fuel processor. It takes around 10-12 seconds to settle down at a new operating point when a step change is applied. As a result, its U_f depicted in Figure 6.10(b) is greater than 0.9 and it is less than 0.7 for the load change of 0-1 p.u. and 1-0.5 p.u. respectively. According to (6.7), with the step change in reference power, I_{sofc} (numerator) will change instantaneously, where as the $f_{H_2}^{in}$ (denominator) is changing slowly, resulting into abrupt and undesirable change in U_f. It crosses its limits of 0.7 to 0.9 at every change considered here. This shows

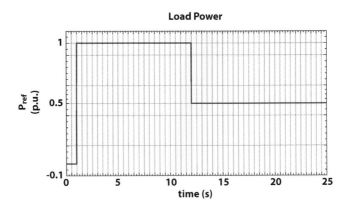

Figure 6.9 The waveform of load power.

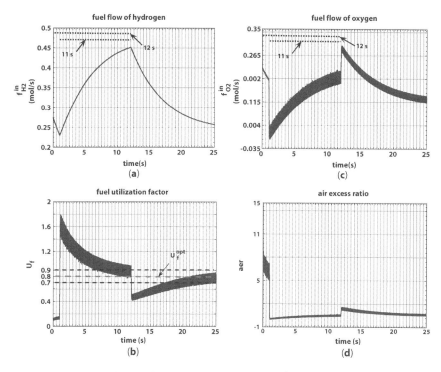

Figure 6.10 The waveforms of (a) fuel flow of hydrogen, $f_{H_2}^{in}$, (b) fuel utilization factor, U_f (c) fuel flow of oxygen, $f_{O_2}^{in}$ and (d) air excess ratio, *aer* in SOFC.

that the SOFC is over and underused during the process in order to satisfy the load change.

Although attempts have been made in (6.9) and (6.10) to generate the fuel feed and the fuel flow at the inlet of SOFC for optimum value of 0.8 of its fuel utilization factor, still it is settling sluggishly and consuming large time of 10-12 seconds.

Further, comparison of Figure 6.10 (a) and (c) reveals that the pattern of the waveforms of $f_{H_2}^{in}$ and $f_{O_2}^{in}$ are the same because according to (6.11), $f_{O_2}^{in}$ is a constant (r_{H_O}) multiple of $f_{H_2}^{in}$. And the variable *aer* depicted in Figure 6.10(d) is showing the reverse pattern of U_f, because the SOFC current and fuel input has been interchanged (refer (6.8)). The explanation of pattern can be understood in the same manner as already explained for U_f.

Therefore, it indicates that the SOFC is a slow responding device and an effective mechanism to avoid the over usage of it during $U_f \geq 0.9$ and under usage during $U_f \leq 0.7$ is required because both these conditions are unsecured operations [25]. It is vital to note here that the load change on an actual power system is very frequent in contrast to the load change

160 ENERGY STORAGE FOR MODERN POWER SYSTEM OPERATIONS

considered in this study. These undesirable conditions will be adverse and will affect the performance of this cell severely. Hence, there exists a strong need to improve the control strategy to enhance the life of the fuel cell by keeping the U_f within its limits.

6.4 Illustration of a Case Study on Control of Grid-Connected SOFC

Apart from the problem observed in the previous section that the utilization factor crosses its limits during transient period, the control problem becomes more complex when this fuel cell is connected to the AC utility grid. This is because SOFC is a slow-responding device (as discussed in the previous section) and the inverter that is essential for converting DC into AC, is a fast-acting device. According to (6.7), U_f can be controlled either by I_{sofc} or $f_{H_2}^{in}$. SOFC current, I_{sofc} can be controlled with inverter, whereas, $f_{H_2}^{in}$ can be controlled with SOFC. The time constant of inverter is very small in comparison to SOFC. Thus, for a smooth operation of grid-connected SOFC, there is a need for proper control of these slow- and fast-acting devices.

While maintaining the coordination between the two devices, it is essential that SOFC stay inside its feasible operational limits irrespective of the steady or transient period of operation. The feasibility of its operation is bounded in terms of both electrical and chemical parameters as given in (6.6). The need of feasibility arises in order to have a better dynamical performance under practical scenario without affecting the life of SOFC during sudden interventions in the system [57]. Since it ages fast due to its thermal cycle limitations, therefore, feasible operational area for SOFC is vital [58, 59]. In addition, the possibility of various losses in the cells such as activation and diffusion losses increases due to unbounded feasible operation. This gradually decreases the overall performance of SOFC operating in a system [19].

Therefore, this section represents the augmented feed forward law to keep the SOFC within its FOA for a grid-connected system to address the above-discussed issues. The scheme based on the augmented feed forward law for a grid-connected SOFC model is shown in Figure 6.11.

Whenever U_f deviates from its optimum value of 0.8, the change in fuel utilization factor, ΔU_f is generated as given in (6.12).

$$U_f = \left| U_f - U_f^{opt} \right| \tag{6.12}$$

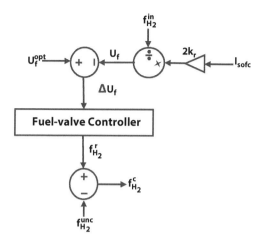

Figure 6.11 A schematic for augmented control scheme of 'constant fuel utilization factor' for SOFC-based power system.

In this control scheme, ΔU_f is regulated with the help of a fuel valve controller that utilizes PI-based controller which has control parameters, proportional and integral constants as $K_p = 5$ and $K_i = 1$ [20].

The control side has recently gained popularity and many researchers have suggested the robust control designs for the SOFCs, such as GPC [60], and NNPC [61] through simulation studies. The accuracy of traditional controllers such as PI is long-term proven due to its high performance under practical scenarios [62, 63]. Thus, it helps in fastening the operation of fuel processing by incorporating a pole in the model to yield the natural fuel flow of hydrogen, $f_{H_2}^r$. As a result, the controlled/refined fuel feed, $f_{H_2}^c$ is obtained as given by (6.13)

$$f_{H_2}^c = \left| f_{H_2}^r - f_{H_2}^{unc} \right| \tag{6.13}$$

Therefore, through a unity feedback loop of $f_{H_2}^{in}$ in the fuel processor [20], (6.10) is represented again with this control scheme in (6.14),

$$f_{H_2}^{in} = \frac{f_{H_2}^c}{1 + s\tau_f} \tag{6.14}$$

Now this augmented control scheme is applied for providing an effective management of the SOFC in terms of a secured life operation. The SOFC, magnitude and time for current change, simulation time are the same as that of the previous model. After running the model with the augmented control scheme, the response of variables associated with SOFC, i.e., $f_{H_2}^{in}$ and $f_{O_2}^{in}$, U_f and *aer* have been shown in Figure 6.12.

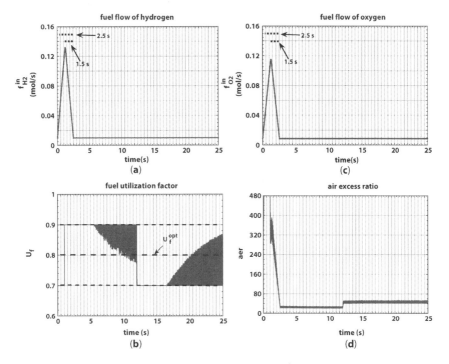

Figure 6.12 The waveforms of (a) fuel flow of hydrogen, $f_{H_2}^{in}$ (b) fuel flow of oxygen, $f_{O_2}^{in}$ (c) fuel utilization factor, U_f and (d) air excess ratio, aer in SOFC during augmented control.

In Figure 6.12(a), it is seen that that this augmented control strategy settles $f_{H_2}^{in}$ within 1.5-2.5 seconds in contrast to 10-12 seconds with previous control strategy. It is due to the incorporation of a fuel value controller that yields $f_{H_2}^r$ according to (6.13); as a result, the controlled fuel is thirsted towards the fuel processor. Thereby as per (6.14) SOFC inlet gets a controlled fuel flow of hydrogen, $f_{H_2}^{in}$. As a result, referring (6.7), it can be seen that, $f_{H_2}^{in}$ is changing at a fast rate, therefore, U_f depicted in Figure 6.12(b) is within the limits of 0.9 to 0.7 irrespective of the considered load power changes. This shows that the SOFC is not being over or underused during the process while satisfying the load change.

Further, referring to (6.11) the instant change in $f_{O_2}^{in}$ as shown in Figure 6.12(c) is depicting the similar trend of the instant rise in $f_{H_2}^{in}$. This fast O_2 feeding to the SOFC allows its better participation into the reaction that raises the requisite amount of *aer* as shown in Figure 6.12(d).

Apart from the improvement in the dynamic responses of SOFC, the electrical output characteristics of a grid-connected SOFC system also depict enhancement as shown in Figure 6.13. The output electrical

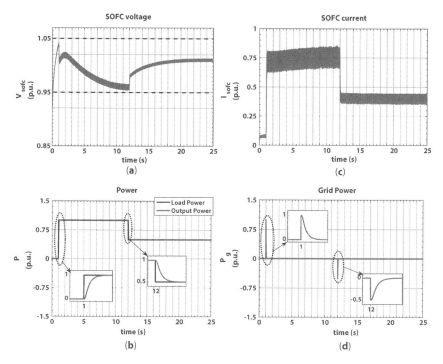

Figure 6.13 The output voltage, V_{sofc}, the SOFC current, I_{sofc}, reference, P_{ref} and generated output power, P_{sofc} and grid power, P_g for constant utilization control operation.

characteristics includes V_{sofc}, I_{sofc}, reference, P_{ref} and generated output power, P_{sofc} and grid power, P_g. During the load change when the need for fuel either arises or drops, V_{sofc} shown in Figure 6.13(a) depicts the same rise/fall characteristics required as per the load change at 1 and 12 seconds (refer to (6.4)). Since the SOFC is able to operate without any risk of being over or underused with this control scheme, a successful operation of SOFC is achieved under feasible region that helps in reducing its aging. This is because the voltage profile is maintained within the FOA as per (6.6) which promotes the secured life for SOFC. In other words, the degradation rate of the cells is lowered that ultimately benefits the system economically.

In Figure 6.13(b), a comparison of P_{ref} and P_{sofc} demonstrates that P_{sofc} follows P_{ref} in less than one second. It indicates fast load tracking capability of SOFC without pushing itself towards over/under usage conditions. It is on the account of SOFC current, I_{sofc} as shown in Figure 6.13(c) which is less than 1 p.u. However, it maintains the FOA limits given in (6.6) and satisfies the feasible operation of SOFC. Finally, Figure 6.13(d) depicts P_g

that indicates less burden of the system on the AC utility grid. Further, Figure 6.14 illustrates V_{abc}, I_{abc} and the frequency $\left(f = \dfrac{2\pi}{\omega}\right)$ at grid.

Comparison of DC and AC voltage profiles in Figures 6.13 and 6.14 reveals that during any load change, there is an appreciable change in V_{sofc}, but at the same time, there is no change in V_{abc} at grid-side. This strengthens the claim that there is no need of voltage control strategy in grid-connected mode of this fuel cell. Additionally, no change is seen on the frequency, f.

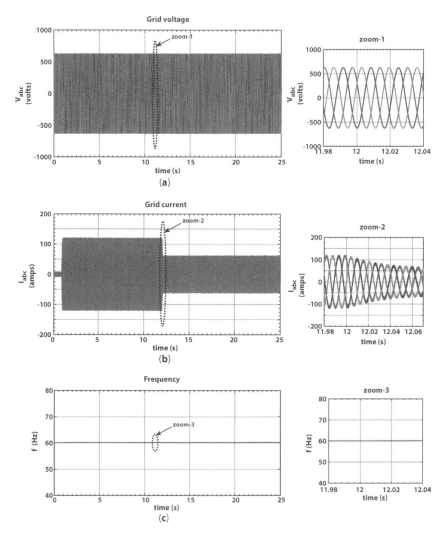

Figure 6.14 The waveforms of (a) AC voltage, (b) AC current and (c) the frequency at grid.

Therefore, constant voltage and frequency during such an appreciable load change is attained which is the first and foremost requirement of any grid-connected renewable energy–based power plant as per electricity grid code.

A systematic development of a coordinated control strategy developed after linearising this non-linear fuel cell depicts that the fuel feed of hydrogen, $\Delta f_{H_2}^{unc}$ and the output change in SOFC voltage, ΔV_{sofc} are the proper control variables, and I_{sofc} is the disturbance signal. With this strategy, it is possible to keep the utilization factor, U_f within limits of 0.7 to 0.9 during transient period, i.e., during any load change, resulting in increased life of this fuel cell. One of the important parameters, i.e., air excess ratio has also been incorporated for analyzing the complete dynamic behavior of this non-linear fuel cell. In addition to this, SOFC is able to meet the instantaneous load change while maintaining the voltage and frequency at one p. u. that shows the enhanced performance of the control strategy applied onto the static power plant under consideration.

6.5 Recent Trend in Fuel Cell Technologies

The futuristic growth of fuel cell technologies entirely depends on the improvements in its fuel energy infrastructure and the rate of decline in other polluting energies such as oil-based power stations [33]. The major growth is in the stationary use of fuel cells such as small-scale grid-connected power generations, uninterruptible power supplies, and small power units at residential places. The fuel cells that are utilized for the stationary application are PEMFC, PAFC, MCFC and SOFC [64]. Among these cells, SOFC is growing very fast due to its inherent advantages and PEMFC has displayed a declining trend over the years as shown in Figure 6.15 with the help of red trend line. Other fuel cells, i.e., PAFC and MCFC have potential to grow but still their usage has not grown significantly in comparison to SOFC. A linear forecast of SOFC and PEMFC as highlighted with the help of an arrow marked in black color that their use in the market is most likely to increase in future [9, 65].

However, the overall commercialization of the fuel cells is not yet on a massive rate and have wide scope for growth enhancement in other applications such as portable power for example, torches, battery chargers and small personal electronics like cameras. In addition, fuel cells such as SOFC can be used for military applications like portable soldier-borne power and skid-mounted generators [9, 64].

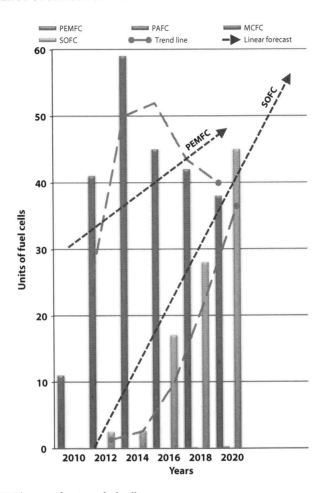

Figure 6.15 The use of various fuel cells.

Therefore, it is vital to penetrate the techno-economic aspect of fuel cells in order to increase their commercialization on a larger scale for these applications as well.

6.5.1 Techno-Economic Comparison

As a new source of power generation, a techno-economic comparison with the batteries and traditional ICEs will open new ways for it [66]. This comparison is built on the following terms:

- the average rate of efficiency provided by the system,

- the rate in the reduction of emissions (carbon in nature) produced by the power source,
- the ease of operation, i.e., during high efficiency requirements, whether the system is easily modified,
- the environmental friendliness such as low noise, air, etc., pollution-levels,
- the life span of the device
- the maintenance cost that includes the costing of associated systems during run or operating condition of the system, and
- the infrastructural requirements.

In Table 6.3, the comparative analysis demonstrates that the average efficiency of fuel cells is twice that of conventional engines and the same as that of batteries. However, batteries can only work for a limited time, i.e., fuel cells can work for longer operating hours in comparison to batteries. Next, the reduction in emission rate is highest by the fuel cells. It is because fuel cells can operate on a wider range of temperatures and emit greenhouse gases. Further, fuel cells are environmental-friendly power sources as they operate silently and provide a smooth operation, therefore no noise pollution. Also, fuel cells have better power supply life in comparison to the batteries [67]. These fuel cells do not require recharging unlike the storage devices such as batteries that generate reactants over a period during recharging processes; as a result there exists a requirement of external

Table 6.3 Techno-economic comparison.

S. no.	Terms	ICEs	Batteries	Fuel cells
1	Average efficiency	20-35%	50-60%	50-60%
2	Emission reduction rate	low	high	very high
3	Ease of operation	complex	partially complex	easy
4	Environmental friendliness	partially	partially	fully
5	Life span	moderate	moderate	very high
6	Maintenance requirement	moderate	moderate	low
7	Infrastructure requirement	moderate	moderate	high

power source in batteries. Thus, fuel cells provide ease of operation because problems like recharging the batteries do not exist, while operating at minimal maintenance cost [3, 68].

However, fuel cells may also lead to high maintenance cost due to unexpected damages or repair requirements in its cells [68]. And the infrastructural requirements by the fuel cells is initially high because these sources of power are upcoming in the system. Concisely, the overall performance of fuel cells is appreciably better than conventional engines and batteries. Therefore, the use of such efficient and environmental-friendly power source could revolutionize the market of power generation.

6.5.2 Market and Policy Barriers

To commercialize in the market, the power source must be reliable, durable, and economical. In addition to it, the feasibility, quality, and user-satisfaction are also silent requirements from the power source by the market [66]. Although the future of fuel cells seems bright in replacing the traditional fossil fuels, it is still facing a few challenges due to less work in the R&D sectors. There exist a scope for enhancing the durability and reliability of fuel cells, especially in the area of structures of its components, such as catalyst layer and dimensions, material stress, i.e., assessment of the new properties, such as porosity and system modeling and control [68]. Therefore, fuel cell discussed in detail, i.e., SOFC, which is efficiently reliable on distribution networks [11–13], has become more durable with the improvement in its control process as discussed in section 3. With its increased life security, it ultimately translates reduced operating cost for the same power generation capacity in comparison to the conventional engines [66]. Therefore, it becomes more economical in nature. Further, the load tracking capability defines its quality during the load change when the demand from the end user fluctuates; this is also well established as shown in Figure 6.13. As a result, it is providing quality power supply with user satisfaction.

Due to the similar R&Ds, in the last decade fuel cells have created a place in the global market. These are used in California, Texas and Pennsylvania as backup power at telecommunications sites and in forklifts for materials handling [64]. In India a high-pressure hydrogen storage and dispensing terminal for fuel cell–based vehicles with the refuelling station located in New Delhi implemented by Indian Oil Corporation Limited (IOCL) is already in working condition [10]. Due to favorable climatic conditions and user demand in India, fuel cell–based power generations are beneficial for the water irrigation and telecommunications in the off-grid areas. Thus,

India is a booming market for fuel cells and it has set its goals for the rapid expansion of the hydrogen economy with policies like "Make in India" [69]. Similarly, across the globe many projects and policies will mature; London, Paris, Madrid and Athens will only buy emission-free vehicles for their 9,500 fleet, and European cities are expected to reduce the use of conventional fossil-fuel vehicles to half in the near future [71].

However, market policies are the barriers to the growth of fuel cells ahead of the conventional power sources. The market structure in many parts of the world, such as Europe, is completely commercially deployed and is complex. High electricity price is the biggest difficulty in expanding the market for fuel cells, for example, in Germany where many power stations have been built.

6.6 Summary and Future Scope

The distributed generation system comprising intermittent RES and conventional BESS, connected with grid gives satisfactory results. However, the BESS suffers from problems like generation of reactants due to recharging and it provides limited power supply. In contrast to this, another type of distributed generation system that comprises a non-intermittent renewable source, i.e., SOFC does not suffer from any of these mentioned disadvantages and it can alone serve the purpose of two units (RES and BESS) in a grid-connected system. However, the main hindrance in its commercialization is its cost because it has a short life span. Therefore, this chapter summarizes a step-by-step development of a coordinated control strategy to increase the life of SOFC. The control strategy optimizes the fuel utilization in SOFC within safe limits during a transient period, i.e., during any load change. As a result, there is an increase in its life. Therefore, it means that a substantial reduction in the operating cost occurs for the same power generation capacity. Hence, the implementation of such a control strategy enhances the durability of SOFC along with reduction in cost.

The proposed control strategy is implemented with conventional and well-established PI-based controller. There are other controllers that are based on artificial intelligent techniques such as PSO, Fuzzy and ANFIS. These artificially intelligent controllers are robust and adaptive in nature. Moreover, nowadays these advanced controllers are easily available and exhibit better performance. Therefore, the work can be extended by implementing these controllers for enhancing the dynamic performance of the fuel cell, i.e., to achieve the optimum value of its fuel utilization factor exactly at '0.8' during transient as well as during steady-state period.

Further, the enhancement of load tracking capability and modeling of voltage source inverters associated with slow-acting SOFC are some of the other areas of research. For the sake of completeness, this chapter also gives various types of fuel cells, their advantages, disadvantages, applications and their recent trends in the market.

Acknowledgement

The authors are thankful to MHRD, Govt. of India, for giving the grant to U.I.E.T, Panjab University Chandigarh for Design Innovation Centre (DIC) vide letter no. 17-11/2015-PN.1 for providing the research facilities.

References

1. P. Shen and N. Lior, "Vulnerability to climate change impacts of present renewable energy systems designed for achieving net-zero energy buildings," *Energy*, vol. 114, pp. 1288–1305, 2016.
2. X. Li and S. Wang, "A review on energy management, operation control and application methods for grid battery energy storage systems," *CSEE J. Power Energy Syst.*, Jun. 2019.
3. T. Wilberforce, A. Alaswad, A. Palumbo, M. Dassisti, and A. G. Olabi, "Advances in stationary and portable fuel cell applications," *Int. J. Hydrogen Energy*, vol. 41, no. 37, pp. 1–14, Oct. 2016.
4. S. Srinivasan, *Fuel Cells: From Fundamentals to Applications*. Springer, US, 2006.
5. M. Alhassan and M. U. Garba, "Design of an Alkaline Fuel Cell," *Leonardo Electron. J. Pract. Technol.*, no. 9, pp. 99–106, 2006.
6. G. Kaur, *Solid Oxide Fuel Cell Components*. Springer International Publishing, 2016.
7. N. H. Behling, *Fuel Cells: Current Technology Challenges and Future Research Needs*. Newnes, 2012.
8. D. Hart, F. Lehner, R. Rose, and J. Lewis, "The Fuel Cell Industry Review 2014," pp. 1–52, 2014.
9. D. Hart, F. Lehner, R. Rose, and J. Lewis, "The Fuel Cell Industry Review 2018." pp. 1–52, 2018.
10. L. A. Turner and D. A. Mathur, "Driving India towards the clean energy technology frontier." pp. 1–12, 2019.
11. P. Costamagna, A. De Giorgi, L. Magistri, G. Moser, L. Pellaco, and A. Trucco, "A Classification Approach for Model-Based Fault Diagnosis in Power Generation Systems Based on Solid Oxide Fuel Cells," *IEEE Trans. Energy Convers.*, vol. 31, no. 2, pp. 676–687, 2016.

12. J. Milewski, M. Wołowicz, and J. Lewandowski, "Comparison of SOE/SOFC system configurations for a peak hydrogen power plant," *Int. J. Hydrogen Energy*, vol. 42, no. 5, pp. 3498–3509, 2017.
13. P. Sarmah and T. K. Gogoi, "Performance comparison of SOFC integrated combined power systems with three different bottoming steam turbine cycles," *Energy Convers. Manag.*, vol. 132, pp. 91–101, 2017.
14. A. Buonomano, F. Calise, M. D. d'Accadia, A. Palombo, and M. Vicidomini, "Hybrid solid oxide fuel cells-gas turbine systems for combined heat and power: A review," *Appl. Energy*, vol. 156, pp. 32–85, 2015.
15. K. Fischer and J. R. Seume, "Thermo-mechanical stress in tubular solid oxide fuel cells: Part II – Operating strategy for reduced probability of fracture failure," *IET Renewable Power Generation*, vol. 6, no. 3, p. 194, 2012.
16. Y. Zhu and K. Tomsovic, "Development of models for analyzing the load-following performance of microturbines and fuel cells," *Electr. Power Syst. Res.*, vol. 62, no. 1, pp. 1–11, 2002.
17. D. Marra, C. Pianese, P. Polverino, and M. Sorrentino, *Models for Solid Oxide Fuel Cell Systems*, 1st ed. London: Springer London, 2016.
18. C. Wachter, R. Lunderstadt, and F. Joos, "Dynamic model of a pressurized SOFC/gas turbine hybrid power plant for the development of control concepts," *J. Fuel Cell Sci. Technol.*, vol. 3, no. 3, pp. 271–279, 2006.
19. C. Wang and H. Nehrir, "A Physically Based Dynamic Model for Solid Oxide Fuel Cells," *IEEE Trans. Energy Convers.*, vol. 22, no. 4, pp. 887–897, Jun. 2007.
20. L. Sun, G. Wu, Y. Xue, J. Shen, D. Li, and K. Y. Lee, "Coordinated Control Strategies for SOFC Power Plant in a Microgrid," *IEEE Trans. Energy Convers.*, vol. 33, no. 1, pp. 1–9, 2018.
21. S. A. Taher and S. Mansouri, "Optimal PI controller design for active power in grid-connected SOFC DG system," *Int. J. Electr. Power Energy Syst.*, vol. 60, pp. 268–274, Sep. 2014.
22. Y. H. Li, S. S. Choi, and S. Rajakaruna, "An analysis of the control and operation of a solid oxide fuel-cell power plant in an isolated system," *IEEE Trans. Energy Convers.*, vol. 20, no. 2, pp. 381–387, 2005.
23. Y. H. Li, S. Rajakaruna, and S. S. Choi, "Control of a solid oxide fuel cell power plant in a grid-connected system," *IEEE Trans. Energy Convers.*, vol. 22, no. 2, pp. 405–413, 2007.
24. J. Lee, J. Jo, S. Choi, and S. B. Han, "A 10-kW SOFC low-voltage battery hybrid power conditioning system for residential use," *IEEE Trans. Energy Convers.*, vol. 21, no. 2, pp. 575–585, 2006.
25. J. H. Yi and T. S. Kim, "Effects of fuel utilization on performance of SOFC/gas turbine combined power generation systems," *J. Mech. Sci. Technol.*, vol. 31, no. 6, pp. 3091–3100, 2017.
26. S. Yu, T. Fernando, T. K. Chau, and H. H. Iu, "Voltage Control Strategies for Solid Oxide Fuel Cell Energy System Connected to Complex Power

Grids Using Dynamic State Estimation and STATCOM," *IEEE Trans. Power Syst.*, vol. 32, no. 4, pp. 3136–3145, 2017.
27. N. R. Merritt, C. Chakraborty, P. Bajpai, and B. C. Pal, "A Unified Control Structure for Grid Connected and Islanded Mode Operations of Voltage Source Converter based Distributed Generation Units under Unbalanced and Non-linear Conditions," *IEEE Trans. Power Deliv.*, vol. 8977, pp. 1–12, 2019.
28. L. Carrette, K. A. Friedrich, and U. Stimming, "Fuel Cells - Fundamentals and Applications," *Fuel Cells*, vol. 1, no. 1, pp. 5–39, 2001.
29. M. W. Ellis, M. R. Von Spakovsky, and D. J. Nelson, "Fuel Cell Systems: Efficient, Flexible Energy Conversion for the 21st Century," *Proc. IEEE*, vol. 89, no. 12, pp. 1808–1817, 2001.
30. M. Turco, A. Ausiello, and L. Micoli, *Treatment of Biogas for Feeding High Temperature Fuel Cells: Removal of Harmful Compounds by Adsorption Processes*. Springer, 2016.
31. P. Costamagna, P. Costa, and V. Antonucci, "Micro-modelling of solid oxide fuel cell electrodes," *Electrochim. Acta*, vol. 43, no. 3–4, pp. 375–394, 1998.
32. N. P. Brandon and F. Riddoch, "DOE State of the States: Fuel Cells in 2016 Report," 2016.
33. S. Basu, "Future Directions of Fuel Cell Science and Technology," in *Recent Trends in Fuel Cell Science and Technology*, no. 1, New York, NY: Springer New York, 2007, pp. 356–365.
34. S. Park, J. M. Vohs, and R. J. Gorte, "Direct oxidation of hydrocarbons in a solid-oxide fuel cell," *Nature*, vol. 404, no. 6775, pp. 265–267, Mar. 2000.
35. T. Hong, Z. Geng, K. Qi, X. Zhao, J. Ambrosio, and D. GU, "A Wide Range Unidirectional Isolated DC-DC Converter for Fuel Cell Electric Vehicles," 2020.
36. A. A. Amamou, M. Kandidayeni, S. Kelouwani, and L. Boulon, "An Online Self Cold Startup Methodology for PEM Fuel Cells in Vehicular Applications," *IEEE Trans. Veh. Technol.*, vol. 9545, pp. 1–13, 2020.
37. J.-M. Kwon, Y.-J. Kim, and H.-J. Cho, "High-Efficiency Active DMFC System for Portable Applications," *IEEE Trans. Power Electron.*, vol. 26, no. 8, pp. 2201–2209, Aug. 2011.
38. Yi Zhang, Jian Lu, Haoshen Zhou, T. Itoh, and R. Maeda, "Application of Nanoimprint Technology in MEMS-Based Micro Direct-Methanol Fuel Cell (μ-DMFC)," *J. Microelectromechanical Syst.*, vol. 17, no. 4, pp. 1020–1028, Aug. 2008.
39. A. Buonomano, F. Calise, M. D. D'Accadia, A. Palombo, and M. Vicidomini, "Hybrid solid oxide fuel cells–gas turbine systems for combined heat and power: A review," *Appl. Energy*, vol. 156, pp. 32–85, Oct. 2015.
40. A. Bertei, J. Mertens, and C. Nicolella, "Electrochemical simulation of planar solid oxide fuel cells with detailed microstructural modeling," *Electrochim. Acta*, vol. 146, pp. 151–163, 2014.

41. A. Gebregergis, P. Pillay, D. Bhattacharyya, and R. Rengaswemy, "Solid Oxide Fuel Cell Modeling," *IEEE Trans. Ind. Electron.*, vol. 56, no. 1, pp. 139–148, 2009.
42. H. Cao, X. Li, Z. Deng, J. Li, and Y. Qin, "Thermal management oriented steady state analysis and optimization of a kW scale solid oxide fuel cell stand-alone system for maximum system efficiency," *Int. J. Hydrogen Energy*, vol. 8, 2013.
43. H. Cao and X. Li, "Thermal Management-Oriented Multivariable Robust Control of a kW-Scale Solid Oxide Fuel Cell," *IEEE Trans. Energy Convers.*, vol. 31, no. 2, pp. 596–605, 2016.
44. L. Barelli, G. Bidini, G. Cinti, and A. Ottaviano, "Study of SOFC-SOE transition on a RSOFC stack," *Int. J. Hydrogen Energy*, vol. 42, pp. 26037–26047, 2017.
45. C. J. Hatziadoniu, A. A. Lobo, F. Pourboghrat, and M. Daneshdoost, "A simplified dynamic model of grid-connected fuel-cell generators," *IEEE Trans. Power Deliv.*, vol. 17, no. 2, pp. 467–473, Apr. 2002.
46. M. D. Lukas, K. Y. Lee, and H. Ghezel-Ayagh, "Development of a stack simulation model for control study on direct reforming molten carbonate fuel cell power plant," *IEEE Trans. Energy Convers.*, vol. 14, no. 4, pp. 1651–1657, 1999.
47. P. Thounthong, A. Luksanasakul, P. Koseeyaporn, and B. Davat, "Intelligent model-based control of a standalone photovoltaic/fuel cell power plant with supercapacitor energy storage," *IEEE Trans. Sustain. Energy*, vol. 4, no. 1, pp. 240–249, 2013.
48. J. Padullés, G. Ault, and J. McDonald, "An integrated SOFC plant dynamic model for power systems simulation," *J. Power Sources*, vol. 86, no. 1–2, pp. 495–500, Mar. 2000.
49. Y. Zhu and K. Tomso, "Development of models for analyzing the load-following performance of microturbines and fuel cells," vol. 62, 2002.
50. Z. Yang, R. Ma, S. Cheng, and M. Zhan, "Nonlinear Modeling and Analysis of Grid-connected Voltage Source Converters Under Voltage Dips," *IEEE J. Emerg. Sel. Top. Power Electron.*, vol. 14, no. 8, pp. 1–12, 2015.
51. M. Hossain, H. Pota, W. Issa, and M. Hossain, "Overview of AC Microgrid Controls with Inverter-Interfaced Generations," *Energies*, vol. 10, no. 9, pp. 1–27, Aug. 2017.
52. S. Habib, M. M. Khan, F. Abbas, A. Ali, and M. T. Faiz, "Contemporary trends in power electronics converters for charging solutions of electric vehicles," *CSEE J. Power Energy Syst.*, pp. 1–36, 2020.
53. D. Ma, W. Chen, L. Shu, X. Qu, X. Zhan, and Z. Liu, "A Multiport Power Electronic Transformer Based on Modular Multilevel Converter and Mixed-frequency Modulation," *IEEE Trans. Circuits Syst. II Express Briefs*, pp. 1549–7747, 2019.
54. A. K. Sahoo, K. Mahmud, M. Crittenden, J. Ravishankar, S. Padmanaban, and F. Blaabjerg, "Communication-less Primary and Secondary Control

in Inverter-Interfaced AC Microgrid: An Overview," *IEEE J. Emerg. Sel. Top. Power Electron.*, vol. 6777, pp. 1–12, 2019.
55. M. A. Hossain, M. I. Azim, M. A. Mahmud, and H. R. Pota, "Primary voltage control of a single-phase inverter using linear quadratic regulator with integrator," in *Australasian Universities Power Engineering Conference (AUPEC)*, 2015, pp. 1–6.
56. H. S. Chung, H. Wang, F. Blaabjerg, and M. Pechi, *Reliability of Power Electronic Converter Systems*, vol. 3, no. 4. 2016.
57. P. Gupta, V. Pahwa, and Y. P. Verma, "Performance enhancement of a grid-connected solid-oxide fuel cell using an improved control scheme," *Mater. Today Proc.*, vol. 28, pp. 1990–1995, 2020.
58. A. Hagen, R. Barfod, P. V. Hendriksen, Y.-L. Liu, and S. Ramousse, "Degradation of Anode Supported SOFCs as a Function of Temperature and Current Load," *J. Electrochem. Soc.*, vol. 153, no. 6, p. A1165, 2006.
59. J. Kim, D. Shin, N. I. Cho, B. Kang, and N. Chang, "Aging management using a reconfigurable switch network for arrays of nonideal power cells," *IEEE Trans. Very Large Scale Integr. Syst.*, vol. 26, no. 5, pp. 855–866, 2018.
60. Z. Deng, H. Cao, X. Li, J. Jiang, J. Yang, and Y. Qin, "Generalized predictive control for fractional order dynamic model of solid oxide fuel cell output power," *J. Power Sources*, vol. 195, no. 24, pp. 8097–8103, 2010.
61. S. A. Hajimolana, M. A. Hussain, W. M. A. Wan Daud, and M. H. Chakrabarti, "Neural network predictive control of a SOFC fuelled with ammonia," *Int. J. Electrochem. Sci.*, vol. 7, no. 4, pp. 3737–3749, 2012.
62. G. M. Pelz, S. A. O. da Silva, and L. P. Sampaio, "Comparative analysis involving PI and state-feedback multi-resonant controllers applied to the grid voltage disturbances rejection of a unified power quality conditioner," *Int. J. Electr. Power Energy Syst.*, vol. 115, no. April 2019, p. 105481, Feb. 2020.
63. M. Jain, A. Rani, N. Pachauri, V. Singh, and A. P. Mittal, "Design of fractional order 2-DOF PI controller for real-time control of heat flow experiment," *Eng. Sci. Technol. an Int. J.*, vol. 22, no. 1, pp. 215–228, Feb. 2019.
64. N. L. Garland, D. C. Papageorgopoulos, and J. M. Stanford, "Hydrogen and fuel cell technology: Progress, challenges, and future directions," *Energy Procedia*, vol. 28, pp. 2–11, 2012.
65. N. Duan, "When Will Speed of Progress in Green Science and Technology Exceed that of Resource Exploitation and Pollutant Generation?," *Engineering*, vol. 4, no. 3, p. 299, 2018.
66. J. Wang, H. Wang, and Y. Fan, "Techno-Economic Challenges of Fuel Cell Commercialization," *Engineering*, vol. 4, no. 3, pp. 352–360, Jun. 2018.
67. A. Telli and S. Barkat, "Distributed grid-connected SOFC supporting a multilevel dynamic voltage restorer," *Energy Syst.*, vol. 10, no. 2, pp. 461–487, May 2018.
68. J. Wang, "System integration, durability and reliability of fuel cells: Challenges and solutions," *Appl. Energy*, vol. 189, pp. 460–479, Mar. 2017.

69. T. Pfeifer *et al.*, "Development of a SOFC Power Generator for the Indian Market," *Fuel Cells*, vol. 17, no. 4, pp. 550–561, Aug. 2017.
70. A. Ajanovic and R. Haas, "Economic and Environmental Prospects for Battery Electric- and Fuel Cell Vehicles: A Review," *Fuel Cells*, vol. 19, no. 5, pp. 515–529, Oct. 2019.
71. A. Pramuanjaroenkij, X. Y. Zhou, and S. Kakaç, "Numerical analysis of indirect internal reforming with self-sustained electrochemical promotion catalysts," *Int. J. Hydrogen Energy*, vol. 35, no. 13, pp. 6482–6489, Jul. 2010.
72. S. K. Kamarudin, F. Achmad, and W. R. W. Daud, "Overview on the application of direct methanol fuel cell (DMFC) for portable electronic devices," *Int. J. Hydrogen Energy*, vol. 34, no. 16, pp. 6902–6916, Aug. 2009.

7

Lithium-Ion vs. Redox Flow Batteries – A Techno-Economic Comparative Analysis for Isolated Microgrid System

Maninder Kaur[1]*, Sandeep Dhundhara[2], Sanchita Chauhan[1] and Mandeep Sharma[3]

[1]*Dr. S.S.B. University Institute of Chemical Engineering & Technology, Panjab University, Chandigarh, India*
[2]*Department of Basic Engg., College of Agricultural Engg. and Tech., CCS Haryana Agricultural University, Hisar, India*
[3]*Department of Electrical Engineering, Baba Hira Singh Bhattal Institute of Engineering and Technology, Lehragaga, Punjab, India*

Abstract

The concern over greenhouse gas (GHG) emissions resulting from coal-based thermal power plants has emphasized the role of microgrid systems with locally available renewable energy resources in the energy mix globally. Renewable energy resources cannot produce consistent and reliable power according to the ever-fluctuating load demand owing to their intermittent nature. Thus, storing energy generated from renewable resources for subsequent deployment in the needed hours is the most viable solution to tackle this issue. Various energy storage techniques are available currently to have optimum power storage based on factors such as cost, the lifetime of the battery, storage capacity, response time, and efficiency. This paper focuses on a comparison between the response of two electrochemical batteries; Li-ion and redox flow battery in PV/Biomass isolated microgrid system. The performance of the batteries has been compared considering real prices of components, real resource data, and realistic load demand of the village in Punjab, India.

Keywords: Battery, microgrid, optimization, renewable energy resource, biomass

---

*Corresponding author: maninderkaur2780@gmail.com

Sandeep Dhundhara and Yajvender Pal Verma (eds.) *Energy Storage for Modern Power System Operations*, (177–198) © 2021 Scrivener Publishing LLC

7.1 Introduction to Battery Energy Storage System

Energy storage plays a crucial role in the modern power system as energy created at one time can be preserved for further use at another time enhancing the reliability of the power system. Electrical energy cannot be stored in the form of electricity but can be converted into electrical energy from other forms of energy, such as thermal, mechanical, chemical, magnetic energy through pumped hydro storage system, battery energy storage system (BESS), fuel cells, and superconducting magnetic energy storage system for further use [1, 2]. Among these technologies, BESS is one of the recognized energy storage technologies used to stabilize the electrical networks constituted with highly variable renewable energy resources. Batteries store electrochemical energy in a set of multiple cells, connected in several series/parallel connection individually or both series-parallel connection to achieve desired voltage and current. Lead-acid (LA) battery, one of the matured BESS, which has been conducted for over 150 years, was found to be the most economical energy storage device [3, 4]. For instance, various researchers used lead-acid battery storage systems as a back-up in most isolated or grid-connected microgrids systems.

However, the deprived performance of LA batteries at high and low ambient temperature, and its relatively short life span means that they need to be replaced after every 4 to 5 years. Further, the maintenance of flooded batteries with the periodic supply of water (in flooded battery) and its low specific energy and power rating has led the researchers to look for alternate electrochemical energy storage technologies [5, 6]. Nowadays, Lithium-ion (Li-ion) and redox flow batteries (RFBs) are available in the market as rival technologies to be utilized as back-up storage in the modern power system.

7.1.1 Lithium-Ion Battery

Li-ion batteries (LIBs) technology, developed commercially in the early 1990s is endowed with many attractive properties such as low weight, long cycle life, high energy density, cost, cell voltage, and rate capability. Therefore, LIBs are currently widely used in a variety of applications, such as information technology, aerospace, electric and hybrid vehicles, etc., and most likely to be efficiently utilized to store electric energy in the modern power system. Generally, LIBs are composed of anode and cathode electrodes, filled with liquid electrolyte [7]. The electrochemical reaction takes place between anolytic and catholytic active materials.

i. Cathode:
LIBs are classified into different categories based on the Li-ion donor used as cathode active material such as: lithium iron phosphate (LFP), lithium manganese oxide (LMO), lithium cobalt oxide (LCO), lithium nickel cobalt aluminum (NCA), and lithium nickel manganese cobalt (NMC). Battery performance also significantly differs based on the aforesaid different cathode material. These compounds exhibit high impedance because of their low ionic conductivities and diffusion coefficients which results in low energy efficiency and lifetime [7]. The fabrication of cathode with finely powdered Li compound material and blending with conductive material by mixing with a binder (polyvinylidene fluoride) and a solvent helps to overcome this limitation.

ii. Anode:
The anode in Li-ion cell is composed of carbon-based material such as graphite. The anode active material in Li-ion cell should have the capability to store a large amount of Li-ion and release them easily without damaging the active material. Graphite is the most abundantly available, easily processable, and low-cost carbon material that is most commonly used as an anode electrode in Li-ion cells. It also shows good reversibility on intercalation and deintercalation and forms a stable SEI layer on the first discharge/charge life cycle. The development of different anode materials such as carbon/silicon composite results in significant improvement in the lifetime and capacity of LIBs. Commonly, anode and cathode are referred to as active materials in Li-ion cells. The electrolyte, the separator, current collecting foils, binders, solvents, and other safety devices are referred to as inactive materials [7, 8]. And these inactive materials are required for the operation of the cell without any contribution to the energy storage reactions. For instance, binders mostly made of soft black carbon act as glue which holds the active material together and helps to create a good path for electrical conduction.

iii. Electrolyte:
Electrolyte plays a vital role in electrochemical batteries as it provides the medium to pass the Li-ion back and forth between active material. An electrolyte is basically a solution in which organic solvents are mixed with solute made up of lithium salts and other additives. The electrolyte in Li-ion cells is usually a non-aqueous organic liquid such as ethylene carbonate (EC), dimethyl carbonate (DMC), diethyl carbonate (DEC), ethyl methyl carbonate (EMC), propylene carbonate (PC), consisting of dissolved lithium hexafluorophosphate ($LiPF_6$) salt into it [8].

iv. Separator:
The separator, also known as an ion-conducting membrane, is an ultrathin porous membrane that keeps positive and negative electrodes physically apart, thus preventing a short circuit taking place in the cell. The separator, usually made of extremely thin polymer films like polyolefins, ceramics, not only acts as a safety component within the cell but also must have enough porosity to allow Li-ions to pass back and forth while restricting the flow of electrons [5, 7].

v. Working Principle:
Redox process involving reduction and oxidation reactions takes place in electrochemical Li-ion cell. An oxidation process takes place when an element combines with the more electronegative element and reduction takes place when an element combines with the less electronegative element. Li-ion being cation are positively charged and are always pulled toward more electronegative electrode [6–8]. Therefore, Li-ions are attracted towards the cathode and free electrons move from anode towards cathode causing reduction reaction (gaining of electrons) to occur at the cathode during discharge mode of Li-ion cell as seen in Figure 7.1. The reverse process, oxidation reaction (losing of electrons) occurs at the cathode during charging mode of Li-ion cell as seen from Figure 7.2.

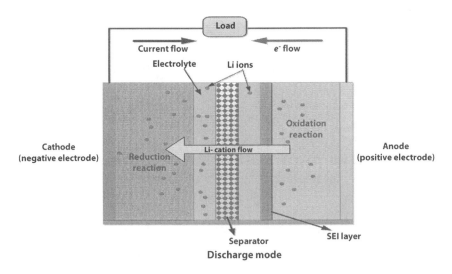

Figure 7.1 Discharging mode of Li-ion cell.

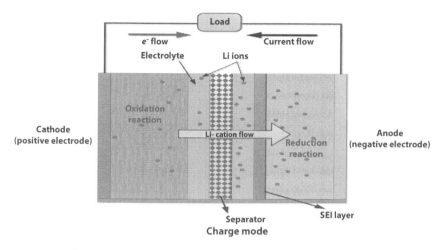

Figure 7.2 Charging mode of Li-ion cell.

vi. Advantages and disadvantages of LIBs:

Li-ion energy storage batteries have wide scope of research in material technology as their size and performance are strongly affected by active materials used for the electrodes and electrolyte. Further, Li-ion batteries have many advantages such as:

- fast charging and discharging capability
- high energy density of 170-300 Wh/l and specific energy 75-125 Wh/kg, respectively [9, 10]
- 78% round-trip efficiency within 3500 cycles.

Li-ion batteries are found to be good candidates for applications where weight and response time are important. In addition to the advantages, these batteries have certain disadvantages like:

- electrolyte used in these batteries is highly inflammable, which raises issues about security and greenness.
- it is harmful when fully discharged [11].
- voltage monitoring system for safe operation, multiple fuse requirement, overcharging safeguard system, higher manufacturing cost, and material make this battery system expensive.
- Lithium being a non-renewable material may cause environmental issues.

In recent years, with shifting of dependency on renewable energy resources to meet the electrical energy demand has also promoted the development of more flexible energy storage systems. Renewable energy resources such as solar and wind are strongly influenced by season and geographical locations. Thus, storing electrical energy during the period of excess generation and supplying the same during hours of need helps to make the renewable energy system–based microgrid system more effective and reliable. LIBs were found to be cost-effective in comparison with lead-acid batteries for different types of microgrid systems by Dhundhara et al. [12]. In another study, Diouf and Pode emphasized the forecasts of LIBs as a major energy storage system in grid system comprising renewable energy sources [13]. Moreover, a company installed a LIB energy storage system with a capacity of 32 MW/8 MWh in New York in 2011 to support the 98 MW wind generation plant [14, 15]. The largest European LIB energy storage with a capacity 6 MW/10 MWh at the primary substation is in process in the United Kingdom (UK) [16]. Recently, Zubi et al. in their study highlighted the promising future of LIBs in the market with the integration of renewable power supply systems such as solar photovoltaics and wind power [17].

7.1.2 Redox Flow Batteries

Redox flow batteries (RFB) were first developed by NASA in 1970. Recently, RFB has been demonstrated for its ability to maintain the load frequency control (LFC), improve power quality, and overcome other operational problems faced by modern power systems. The multiple benefits of RFB such as high energy density, large capacity, ease of recycling, fast response time, and controllable cell temperature over other energy storage batteries favors its use for energy storage application in modern power systems [18, 19]. The method of electrolyte storage in RFB batteries distinguishes it from other electrochemical energy storage systems. The electrolyte in RFB is stored in two external tanks which can be kept away from the battery center [20].

The vanadium redox flow battery (VRFB) is found to be the most promising, commonly researched, and pursued RFB technology. The vanadium ions in four different oxidation states: V^{4+}/V^{5+} and V^{2+}/V^{3+} are stored in two electrolyte tanks as seen from Figure 7.3, such that separate redox couple reaction can take place in each tank. Each cell is composed of two half-cells consisting of an electrode and bipolar plate which are separated by a membrane to allow selective ion exchange while preventing the mixing of electrolytes. The stack is composed of sharing of the bipolar plates of adjacent cells.

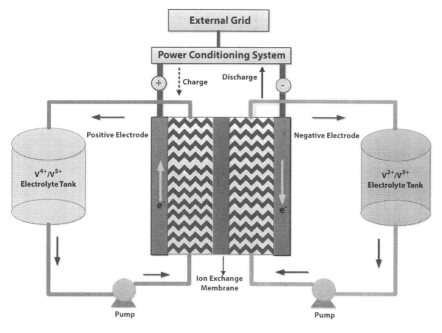

Figure 7.3 Vanadium redox flow battery.

i. Electrolyte:
Electrolytes in VRFB are comprised of an active redox species and supporting solvent or salt. Various vanadium compounds such as: vanadium trichloride (VCl_3), vanadyl sulphate ($VOSO_4$), and vanadium pentoxide (V_2O_5) were considered with sodium hydroxide (NaOH), hydrochloric acid (HCl), and sulfuric acid (H_2SO_4) as supporting electrolyte in VRFB. However, the possibility of chlorine gas formation with the use of vanadium trichloride (VCl_3) and hydrochloric acid (HCl), and low solubility of vanadium pentoxide (V_2O_5) in acids exclude their use as the electrolyte. Thus, vanadyl sulphate ($VOSO_4$) and sulfuric acid (H_2SO_4) were selected initially, though methods for creating electrolytes using vanadium pentoxide (V_2O_5) have been developed afterward [21]. Nowadays, vanadium pentoxide (V_2O_5) is preferred by a group of researchers to reduce the cost of electrolytes while vanadyl sulphate ($VOSO_4$) is preferred when the concentration of vanadium intends to be altered easily.

ii. Membrane:
The membrane is one key component of VRFB that prevents the cross-mixing of vanadium ions while providing ionic conductivity and also affects the economic viability of VRFB significantly. Two types of membranes used in VRFB are ion exchange membrane and a non-ionic porous membrane.

The ion exchange membrane is further classified as an anion, cation, and amphoteric exchange membrane depending on the ionic functional group attached to the membrane. Ideally, the membrane should have low water uptake, high ionic conductivity, good ionic exchange capacity, good chemical and thermal stability, low area electrical resistance, low swelling ratio, low vanadium, and other polyhalide ions permeability as well as low cost [22]. So, the research in recent years has emphasized new technologies and materials to find suitable membranes. Apart from preventing the mixing of electrolyte solution, the membrane also facilitates the redox reactions taking place in RFB by allowing the hydrogen ions to pass, balance, and complete the net reaction.

iii. Working Principle:

The oxidation and reduction reactions within flow batteries are facilitated by electrodes. The electrode surface act as a catalyst and its porous surface as the site for reactions occurring in the electrolyte solution. The reactions taking place on the electrode must be reversible, which requires the oxidation and reduction species to be soluble in the electrolyte solution.

The chemical reaction taking place on the positive side and negative side electrode are as depicted below in equation (7.1) and equation (7.2):

$$\text{Positive cell reaction}: V^{5+} + 2H_2 + e^- \rightleftharpoons V^{4+} + H_2O \quad (7.1)$$

$$\text{Negative cell reaction}: V^{2+} \rightleftharpoons V^{3+} + e^- \quad (7.2)$$

$$\text{Overall cell reaction}: V^{5+} + 2H_2 + V^{2+} \rightleftharpoons V^{4+} + V^{3+} + H_2O \quad (7.3)$$

An array for larger applications can be created by using electrically connected multiple stacks. The electrode material in VRFB is so selected that it must have high corrosion resistivity. As most of the supporting electrolyte material is acidic in nature making the cell environment corrosive [22]. Thus, carbon and graphite material is most often used to form electrodes in VRFB as these materials are good electrical conductors and corrosion resistant to sulfuric acid. Thus, graphitized polyacrylamide (PAN) and cellulose fiber (Rayon) in carbon paper and felt form are the most commonly used electrode material. Similarly, bipolar plates in VRFB are composed of graphite material due to their high conductivity and resistance to corrosion.

iv. Advantages and disadvantages of VRFB:

These unique constructional features of VRFB result in several advantages over other energy storage technologies such as:

- possibility of enhancing the capacity as per the requirement with ease.
- ease of recycling while decommissioning the storage unit.
- the electrolyte solution of the VRFB can be placed in two different tanks which can be put at a distance away from the battery storage unit or underground at any place.
- excellent practicality as it can be installed and distributed in suburban areas.
- no self-discharge.
- independently tunable power rating and energy capacity.
- fast response time and controllable cell temperature.
- fast charging by replacing exhaust electrolyte.
- as the output section (cells) and capacity section (tanks) are independent of each other therefore optimal design depending on application is possible.
- solution contamination with vanadium ion diffusion across the membrane is prevented.
- power conversion and energy storage are independent of each other in RFB.

In contrast, some disadvantages of VRFB technology include:

- high capital cost involved in VRFB energy storage technology.
- cell efficiency adversely affected by gas evolution (produced at electrodes).
- gas evolution also results in reduced electrode surface area and battery charge depletion.
- deterioration of some ion exchange membrane and positive electrode terminal with high oxidation properties of V^{5+}
- mechanical parts such as pumps and flow also require additional maintenance and management.

Thus, VRFB with large availability, long life cycle, low toxicity, and high energy efficiency seems to be one of the most competitive electrochemical battery storage systems. VRFB already has been tested in Italy (0.45 MW/1.44 MW h), the United States (2 MW/8 MW h), and Germany (0.3 MW/1.3 MW h) to date [23]. VRFB in integration with other renewable energy resource systems has the potential to enhance the electric supply capacity in near future [24, 25].

7.2 Role of Battery Energy Storage System in Microgrids

Microgrids comprised of renewable energy resources not only help in meeting the ever-growing electricity demands but also help to diminish the dependency on a fossil fuel–based energy generation system. Thus, electricity generation using locally available renewable energy resources will help to significantly curb GHGs emissions. However, the intermittent nature of renewable energy resources like solar and wind makes them unreliable and inefficient [26].

Consequently, the energy storage system plays a significant role in the microgrid system as it can rapidly adjust the power output, reduce the harmonic components of the system, curbs the frequency and voltage fluctuation, and helps to improve the energy quality of the microgrid. Moreover, power supply interruption can be avoided with the help of BESS in the microgrid through a smooth transition and rapid disconnection when the distribution network fails [8]. In case of voltage fluctuations, the energy storage system can rapidly provide reactive power support to improve regional voltage stability and the energy quality of the microgrid. In a microgrid system composed of solar/wind/battery energy storage system connected to the grid or off-grid mode the state of charge of the battery at any time of the year should meet the following equation requirement:

$$P_{Es,t} \geq \max | P_{L,t} - (P_{WG,t} + P_{PV,t}) | \quad (7.4)$$

where $P_{ES;t}$ represents the rated power of the energy storage battery; $P_{L,t}$ is the load power requirements; $P_{WG,t}$ and $P_{PV,t}$ is the instantaneous power of wind generation and PV generation, respectively.

As of now, lead-acid batteries are most frequently used as BESS in grid-connected or off-grid microgrid designing because of the simplicity of the technology and cost effectiveness. However, although it is a well-established technology, no remarkable improvement has been achieved for a long time. Therefore, future research should emphasize other battery energy storage systems with more potential such as Li-ion and Vanadium redox flow batteries, as the Li-ion battery has been found to be economical in comparison with old and matured technology of lead-acid batteries for different loads. However, VRFB has its own merits to be applicable in modern power system such as the ability to maintain the load frequency control (LFC) and power quality improvement. With the vast scalability and flexibility of VRFB energy storage system, it seems to be an emerging storage technology with the hope to achieve optimal energy storage system in the microgrid system. Based on the literature review, Table 7.1 represents the comparative analysis of Li-ion and VRFB energy storage systems. Although these LIBs and VRFB are relatively young technologies they fulfill all the exemplary traits of an optimal energy storage system.

Table 7.1 Comparison between Li-ion and VRFB energy storage system.

	Metrics	Li-ion	VRFB
Sub-chemistries		LMO, LFP. NCA, NMC, LTO	HCl-based, H_2SO_4-based
Technical parameters	Energy density	170-300 Wh/l	20-33 Wh/l
	Specific energy	75-125 Wh/kg	15-25 Wh/kg
	Discharge time	0.5-5 hours	3-10 hours
	Calendar life	3-10 years	20-25 years
	Depth of discharge	80-95%	100%
	Ambient temperature	0-30 °C	-20-50 °C
	24 hour self-discharge	5%	2.5%
Performance	Degradation	Frequent use, temperature, and deep discharges accelerate the degradation	No degradation and longer life
Safety		Electrolyte spillage is main risk No thermal runaway risk involved	Thermal runaway risk
Capital cost		Capital cost involved lower than RFB	High Capital cost required
Applications		Power and energy storage application	Suitable for energy storage application

Therefore, the study has been conducted to analyze the impact of these two types of batteries in a modern power system by designing a microgrid system to fulfill the load demand of the village.

7.3 Case Study to Investigate the Impact of Li-Ion and VRFB Energy Storage System in Microgrid System

The study has been conducted to optimize the PV/biomass isolated microgrid system with Li-ion battery storage system and redox flow battery energy storage in a modern power system by designing a microgrid system to fulfill the load demand of the village. The isolated microgrid system has been designed using the available renewable resources of the study area and real-time prices of different components used in the microgrid system.

7.3.1 System Modelling

The microgrid system has been designed by taking into consideration the load requirement of Gudana village of Punjab, India. The schematic of a proposed microgrid system depending on the renewable energy resources availability of study area consists of different components such as photovoltaic systems, biogas generator, battery energy storage system, converter and interconnected loads of the consumers, as shown in Figure 7.4. The bidirectional converter maintains the flow of electricity between the AC

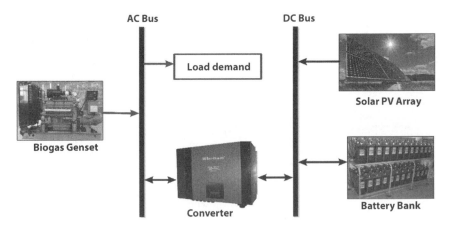

Figure 7.4 Schematic diagram of microgrid system.

and DC buses as it converts DC from PV panels to AC and stores extra energy in the battery.

i. Solar Photovoltaic Generation

A solar-photovoltaic (PV) flat plate, PV panel is taken into consideration with output power (P_{PV}) calculated using equation (7.5):

$$P_{PV} = C_{PV} D_{PV} \left(\frac{I_T}{I_{T,STC}} \right) [1 + \alpha_P (T_C - T_{C,STC})] \qquad (7.5)$$

where C_{PV} denotes the rated capacity of PV in kW and D_{PV} represents the derating factor. I_T represents the solar radiation incident on the PV array in kW/m². $I_{T,STC}$ characterizes the solar radiation incident under standard test conditions in kW/m², α_p is the temperature coefficient of power in %/°C. The derating factor is taken as 80% and the efficiency of PV panels at standard test conditions is taken as 14.5%. The lifetime of the PV panels is taken as 20 years. The 840 $/kW is taken as capital cost while replacement and operation & maintenance (O&M) cost of the PV is taken as 800 $/kW and 10 $/kW/year, respectively with one USD ($) taken as 64 INR [12]. The capital cost includes the charges for delivery, installation, wiring, and mounting hardware.

ii. Biogas Genset

The annual power output of biogas generator set (genset) PA (kWh), can be computed using the following equation (7.6):

$$P_A = P_M \times CUF[365 \times (\text{operating hours/day})] \qquad (7.6)$$

where P_M is the maximum rating and CUF represents the capacity utilization factor of the biogas genset system [27, 28]. The minimum biogas genset of 50 kW was considered for optimization with a lifetime of 20,000 h. The 600 $/kW is taken as capital cost while 450 $/kW and 0.025 $/h are taken as replacement cost and operation & maintenance cost of the biogas genset, respectively [29].

iii. Converter

The rating of the converter (P_{con}) can be expressed by the equation (7.7) as given below:

$$P_{con} = P_L / \eta_{con} \qquad (7.7)$$

where P_L represents the peak load demand and η_{con} denotes the converter efficiency [28–30]. In this study, the life span of 20 years was taken under consideration, with a round trip efficiency of 90%. Here, the capital and replacement cost of the converter has been considered 127 $/kW each. The O&M cost of the converter is taken as 1 $/kW/year [29].

iv. Battery

Batteries play a vital role in enhancing the reliability and stability of the power supply in a microgrid system. Battery energy storage medium is a clean and green technology which stores excess generated energy and discharge as per the requirement of the load demand.

The power rating of the battery can be calculated using the below equation (7.8):

$$P_{bat}^{max} = \frac{N_{bat} V_{bat} I_{bat}^{max}}{1000} \qquad (7.8)$$

where N_{batt}, V_{batt}, and I_{batt} represents represent the number of batteries, voltage rating of a single battery, and maximum charging current rating of the battery in amperes, respectively [12, 29–32]. As Li-ion battery and RFB battery comparative analysis is considered in this study, the different parameters taken into consideration are represented in Table 7.2.

Table 7.2 Technical parameters of batteries.

Parameters	Battery type	
	Li-ion battery	Redox flow battery
Maximum capacity (Ah)	276 Ah	417 Ah
Nominal Capacity	1 kWh	1 kWh
Cycle life @ maximum depth of discharge (DOD)	3200	10000
Float Life (years)	10	20
Initial state of charge (SOC %)	100	100
Minimum state of charge (SOC %)	20	10
Round-trip efficiency (%)	90	80
Cost ($/kWh)	271	452
Replacement cost ($)	200	300
Operation and maintenance cost ($)	11.47	12.02
References	[12, 20, 33]	[20, 33]

7.3.2 Evaluation Criteria for a Microgrid System

The economic feasibility of any microgrid system depends on the total net present cost (TNPC) and per-unit cost of energy (COE) [34, 35]. So, TNPC and COE for the proposed microgrid system are calculated as follows equation (7.9):

$$\text{TNPC} = \frac{C_A}{\text{CRF}(i,n)} \quad (7.9)$$

where C_A represents the total annualized cost ($/year) and CRF represents the capital recovery factor with i as the nominal interest rate (%) and project lifetime represented by n years is estimated as follows equation (7.10):

$$\text{CRF}(i,n) = \frac{i(1+i)^n}{(1+i)^n - 1} \quad (7.10)$$

The COE generated is calculated as the ratio of the annual cost of system components (C_A) to the energy generated. COE can be expressed as equation (7.11):

$$\text{COE} = \frac{C_A}{E_T} \quad (7.11)$$

7.3.3 Load and Resource Assessment

The village load demand mainly constitutes domestic load such as CFLs, fans, television, and refrigerator load with certain specific miscellaneous loads such as mobile charging point, mixer load, and electric iron load. The yearly load profile (yellow shaded area) of the village as seen from Figure 7.5 shows that demand is high during the summer season in the months of June-July as compared to the other seasons throughout the year.

The village has been bestowed with sufficient solar energy with over 300 sunny days in a year, having annual average solar radiation of 5.44 kWh/m²/day, and an average clearness index of 0.633. Agriculture and livestock are the primary sources of income in the village [29, 30]. Therefore, it has sufficient biomass resources in the form of crop residues such as rice, maize, potatoes, etc., and animal manure from buffaloes, cows, pigs, etc., to operate biogas plants. Thus, the village has sufficient biomass energy as a renewable energy resource to become a clean and green self-dependent village.

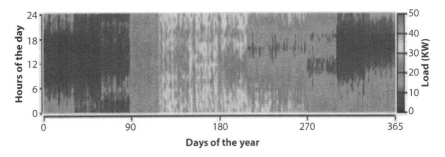

Figure 7.5 Load demand of study area.

7.4 Results and Discussion

The comparative analysis of LIBs and VRFB effect on the performance of PV/biogas/battery connected isolated microgrid system is carried out in this section. The system was developed and investigated using HOMER Pro edition software with load following (LF) dispatch strategy.

The microgrid system using LIBs as a storage system resulted in an optimal system comprised of 50 kW Genset, 61.7 kW PV system, 44.8 kW of system converter, and 68 strings of Li-ion battery. However, a microgrid system utilizing the VRFB as an energy storage system resulted in the optimal combination of 50 kW genset, 62.1 kW of the PV system, 44.3 kW of system converter, and 55 strings of a redox flow battery. Table 7.3 represents the comparative analysis of the cost summary of microgrid systems utilizing LIBs and VRFB.

It has been analyzed that the capital cost involved in a microgrid system with VRFB as a storage system is higher than that of a LIBs storage system. However, the microgrid system with VRFB battery storage system can provide the electricity at 0.129 $/kWh per unit cost lower than that of LIBs battery system as seen from Table 7.3. Further, the operating cost

Table 7.3 Cost summary of microgrid system using LIBs and VRFB.

Battery type	Capital cost ($)	Operating cost ($)	Replacement cost ($)	Salvage ($)	TNPC ($)	COE ($/kWh)
Li-ion	105,959	145,291	109,289	-20,018	480,808.20	0.132
Redox flow	112,637	141,428	95,622	-12,838	472,281.90	0.129

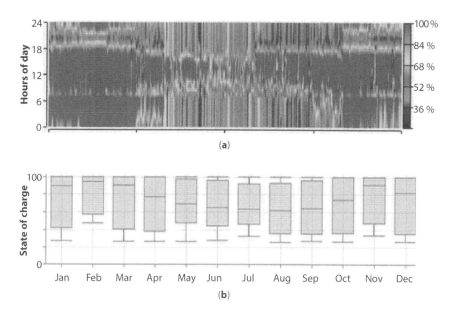

Figure 7.6 State of charge of LIBs (a) Hours of day vs. days of year (b) state of charge vs. months of year.

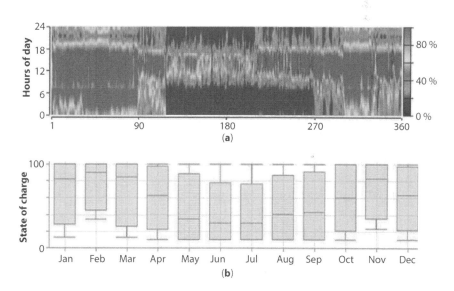

Figure 7.7 State of charge of VRFB (a) Hours of day vs. days of year (b) state of charge vs. months of year.

and replacement cost of a microgrid system with the redox flow battery energy storage system was also found to be smaller than that of LIBs battery energy storage system.

Figure 7.6 and Figure 7.7 represent the state of charge of LIBs and VRFB, respectively, throughout the year. It has been observed from Figure 7.6 that the Li-ion battery remains almost fully charged throughout the year. However, VRFB has been found to be maximum discharged during the peak load period of the summer season as seen from Figure 7.7. Further, it has been observed that VRFB has been utilized to the maximum extent in the microgrid system resulting in a smaller per unit cost of energy.

7.5 Conclusion

The battery energy storage system plays a vital role in the stable and reliable operation of microgrid systems composed of renewable energy resources. Li-ion and Vanadium redox flow batteries energy storage system has immense potential to replace the old and matured lead-acid battery energy storage system in the modern power system. Further, there is wide scope in the improvement of the performance of these latest energy storage technologies with more research in material science. Although it is difficult to analyze the optimal energy storage system in the microgrid system as it depends on various factors such as location, load demand, availability, etc., the comparison between young emerging technologies of Li-ion battery storage system and Vanadium redox flow battery energy has nevertheless been carried out in the present study. The main conclusions of the study are as follows:

- Both energy storage systems are capable of providing the economical per unit cost of electricity with 100% renewable energy system.
- The Vanadium redox flow battery energy storage system was found to be more economical than the Li-ion battery energy storage system.
- The contribution of the Vanadium redox flow battery energy storage system was found to be maximum in the peak load month periods of the year.
- The arising problem of environmental pollution due to the non-decomposition of Lithium on the complete life cycle of Li-ion battery energy storage system can be avoided by utilizing the redox flow batteries.

- Further, the advantage of increasing the capacity of the redox flow battery by increasing the volume of electrolyte solution also favors redox battery in the modern power system. The capacity can be enhanced with an increase in load demand of the study area or as per requirement without causing any damage to the environment.

Thus, the redox flow battery was found to be the most economical energy storage system in the present study and has a wider scope as an energy storage system in the modern power system.

References

1. Dehghani-Sanij AR, Tharumalingam E, Dusseault MB, Fraser R. Study of energy storage systems and environmental challenges of batteries. *Renew Sustain Energy Rev* 2019;104:192–208. doi:10.1016/j.rser.2019.01.023.
2. Mikhail J., Gallego-schmid A., Azapagic A. Environmental sustainability of small-scale biomass power technologies for agricultural communities in developing countries. *Renew Energy*, vol. 141, pp. 493–506, 2019.
3. Diaz-Gonzalez F., Sumper A., Gomis-Bellmunt O., Villafafila-Robles R. A review of energy storage technologies for wind power applications. *Renew Sustain Energy Rev* 2012;16:2154–2171.
4. Alotto P, Guarnieri M, Moro F. Redox flow batteries for the storage of renewable energy: A review. *Renew Sustain Energy Rev* 2014;29:325–35. doi:10.1016/j.rser.2013.08.001.
5. Berrueta A., Martın I., Sanchis P., Ursua A. Lithium-ion batteries as distributed energy storage systems for microgrids. *Distributed Energy Resources in Microgrids* 2019; 143-183. DOI: https://doi.org/10.1016/B978-0-12-817774-7.00006-5
6. Ponce de León C, Frías-Ferrer A, González-García J, Szánto DA, Walsh FC. Redox flow cells for energy conversion. *J Power Sources* 2006;160:716–32.
7. Holger C., Michael S., Kucevic D., Jossen A. Lithium-Ion Battery Storage for the Grid—A Review of Stationary Battery Storage System Design Tailored for Applications in Modern Power Grids. *Energies* 2017;10. doi:10.3390/en10122107
8. Chen T., Jin Y., Lv H., Yang A., Liu M., Chen B., Xie Y., Chen Q. Applications of Lithium Ion Batteries in Grid Scale Energy Storage Systems Trans. of Tianjin University. https://doi.org/10.1007/s12209-020-00236.
9. Lourenssen K., Williams J., Ahmadpour F., Clemmer R., Tasnim S. Vanadium redox flow batteries: A comprehensive review. *J of Energy Storage* 2019;25:100844.

10. Bryans D., Amstutz V., Girault H., Berlouis L. Characterisation of a 200 kW/400 kWh Vanadium Redox Flow Battery. *Batteries* 2018;4:54. doi:10.3390/batteries4040054.
11. Krishan O., Suhag S. An updated review of energy storage systems: Classification and applications in distributed generation power systems incorporating renewable energy resources. *Int J Energy Res* 2018;1–40. doi:10.1002/er.4285.
12. Dhundhara S., Verma Y.P., and Williams A. Techno-economic analysis of the lithium-ion and lead-acid battery in microgrid systems, *Energy Convers Manag* 2018; 177: 122–142.
13. Diouf B., Pode R. Potential of lithium-ion batteries in renewable energy. *Renew Energy* 2015; 76:375–380.
14. Subburaj A.S., Kondur P., Bayne S.B .Analysis and review of grid connected battery in wind applications. In: *2014 6th Annual IEEE Green Technologies Conference* 2014;1–6.
15. Luo X., Wang J.H., Dooner M. Overview of current development in electrical energy storage technologies and the application potential in power system operation. *Appl Energy* 2015; 137:511–536.
16. Taylor P., Bolton R., Stone D. Pathways for energy storage in the UK: a report for the centre for low carbon futures. Centre for Low Carbon Futures, New York. 2012.
17. Zubi G., Dufo-López R., Carvalho M. The lithium-ion battery: state of the art and future perspectives. *Renew Sustain Energy Rev* 2018; 89:292–308.
18. Chakrabarti M., Hajimolana S., Mjalli F., Saleem M., Mustafa I. Redox Flow Battery for Energy Storage. *Arab J Sci Eng* 2013;38:723–39. doi:10.1007/s13369-012-0356-5.
19. Davies T, Tummino J. High-Performance Vanadium Redox Flow Batteries with Graphite Felt Electrodes. *J Carbon Res* 2018;4:1–17. doi:10.3390/c4010008.
20. Redox Flow Batteries. *Energy Stoarge Assoc* 2018:1–3. http://energystorage.org/energystorage/technologies/redox-flow-batteries (accessed July 3, 2018).
21. Shigematru T. Redox Flow Battery for Energy Storage. *Sci Technical Rev* 2011;73:7 doi:10.1007/s13369-012-0356-5.
22. Shia Y., Ezea C., Xiongb B., Hec W., Zhangd H., Lime T., Ukilf A., Zhaoa J. Recent development of membrane for vanadium redox flow battery applications: A review *Appl Energy* 2019;238: 202-224. https://doi. org/10.1016/j.apenergy.2018.12.087
23. Guarnieri M., Trovò A., D'Anzi A., Alotto P. Developing vanadium redox flow technology on a 9-kW 26-kWh industrial scale test facility: design review and early experiments. *Appl Energy* 2018; 230:1425–34.
24. Yang Z., Campanaab P., Yangb Y., Stridhb B., bladc A., Yanab J. Energy flexibility from the consumer: Integrating local electricity and heat supplies in a building. *Appl Energy* 2018; 223:430-442.

25. Arbabzadeh M., Johnson J.X., De Kleine R., Keoleian G.A. Vanadium redox flow batteries to reach greenhouse gas emissions targets in an off-grid configuration. *Appl Energy* 2015; 146:397–408.
26. Nixon J.D., Dey P.K., Davies P.A. The feasibility of hybrid solar-biomass power plants in India. *Energy* 2012;46:541–554.
27. Bhattacharyya S.C. Mini-grid based electrification in Bangladesh: Technical configuration and business analysis. *Renew Energy*, 2015;75.
28. Islam M.S., Akhter R., Rahman M.A. A thorough investigation on hybrid application of biomass gasifier and PV resources to meet energy needs for a northern rural off-grid region of Bangladesh: A potential solution to replicate in rural off-grid areas or not? *Energy* 2018;145: 338–355.
29. Kaur M., Dhundhara S., Verma Y.P., Chauhan S. Techno-economic analysis of photovoltaic-biomass-based microgrid system for reliable rural electrification. *Int Trans. Electri Energy Syst* 2020; 30(5):1–20.
30. J. Ahmad et al., "Techno economic analysis of a wind-photovoltaic-biomass hybrid renewable energy system for rural electrification: A case study of Kallar Kahar," *Energy* 2018.
31. Rajbongshi R., Borgohain D., Mahapatra S. Optimization of PV-biomass-diesel and grid base hybrid energy systems for rural electrification by using HOMER. *Energy* 2017;126: 461–474.
32. Amirkhalili S.A.. Zahedi A.R. Techno-economic Analysis of a Stand-alone Hybrid Wind/Fuel Cell Microgrid System : A Case Study in Kouhin Region in Qazvin 2018; 4:551–560.
33. K Mongird V., Fotedar V., Viswanathan V., Koritarov P., Balducci B., Hadjerioua J., Alam Energy Storage Technology and Cost Characterization Report. *Hydrowires,* U.S. Department of Energy. 2019.
34. Das B.K., Hoque N., Mandal S., Pal T.K., Raihan M.A. A techno-economic feasibility of a stand-alone hybrid power generation for remote area application in Bangladesh. *Energy* 2017; 80: 319-29.
35. Lai C.S., Mcculloch M.D. Levelized cost of electricity for solar photovoltaic and electrical energy storage. *Appl Energy* 2017;190: 191–203.

8

Role of Energy Storage Systems in the Micro-Grid Operation in Presence of Intermittent Renewable Energy Sources and Load Growth

V V S N Murty[1*], Ashwani Kumar[1] and M. Nageswara Rao[2]

[1]*Electrical Department, NIT Kurukshetra, Haryana, India*
[2]*Electrical Department, JNTU Kakinada, Andhra Pradesh, India*

Abstract

Electricity supply systems are shifting towards sustainable and clean energy sources from fossil fuels. Deployment of renewable energy sources (RES) is growing to reduce greenhouse gas emissions and to supply increasing electricity demand. As power output from RES is intermittent in nature, therefore, electricity supply systems shall be planned with more flexible generation to respond effectively to variation in load demand. Hybrid AC/DC micro-grids are the most interesting approaches towards the development of the smart grid concept in the active distribution networks. The advantages of the hybrid micro-grids include higher efficiency, better utilization of renewable, green energy corridor, lower cost of energy, etc. Suitable energy storage system is an essential element of the AC/DC micro-grids because it allows the inclusion of such generation sources within the residential areas easily with no concerns regarding the environmental drawbacks of conventional generation sources. Energy storage systems play a crucial role in maintaining the energy balance and provide reliable power supply in micro-grids integrated with intermittent renewable energy sources.

Keywords: Micro-grid, renewable energy sources, energy storage system, energy management, techno-economic analysis

---

*Corresponding author: murty209@gmail.com

Sandeep Dhundhara and Yajvender Pal Verma (eds.) Energy Storage for Modern Power System Operations, (199–242) © 2021 Scrivener Publishing LLC

8.1 Introduction

Electricity generated in conventional power plants located at remote areas was transferred to load centres through long transmission lines followed by the distribution networks. The conventional distribution network is the passive type, designated to carry power from transmission system to end users. Power flow in passive distribution networks is unidirectional. With the deregulation of electricity markets and environment constraints, applications of distributed generation resources and energy storage devices into the grid are increasing rapidly. Therefore, the traditional passive distribution networks are getting changed to active types with bidirectional power flows.

Evolution of the modern power system from a conventional power system is illustrated in Figure 8.1. Distributed energy resources at the customer premises, BESS and electric vehicles (EV) are indispensable parts of the smart distribution systems. Their participation has brought more dynamics and uncertainties into the grid, and hence new technologies at both planning and operation levels must be developed to manage energy dispatched from distributed energy resources and energy storage units, the charging/discharging of EV so that an entire power distribution system could operate stably and efficiently.

In the low voltage distribution systems, the micro-grid concept in the context of smart grid technology is becoming more popular with the integration of hybrid renewable and energy storage units. Critical requirement

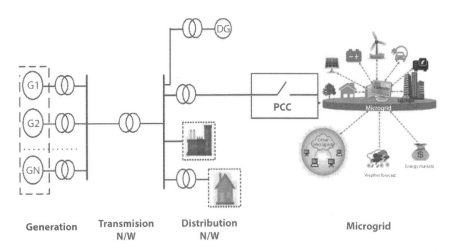

Figure 8.1 Development of modern power system from conventional system.

of stand-alone micro-grid is to maintain the power balance and provide reliable power supply. However, maintaining energy balance is more difficult with an increase in the share of renewables and electric vehicles into the grid, which results in large frequency deviations on the micro-grid. Further, the presence of associated uncertainty factors leads to higher requirements of ancillary services to ensure grid reliability. Battery energy storage systems (BESS) can provide regulating reserve, a type of ancillary service [1].

The Electricity Regulatory Commission (ERC) has to plan for real-time energy markets to achieve maximum savings through optimal energy dispatch with the least cost generation. There is a need for effective energy management strategy in real-time markets that will guide network operators for planning and investment decisions of more flexible generation and energy storage systems, which can respond quickly to variable load demand. Also, ERC needs to plan for adequate capacity of energy storage systems and demand response measures to supplement an increasing share of variable renewable power generation into the grid. It shall also encourage residential, industrial and commercial consumers to participate in demand response and shape electricity consumption profile to take the benefit of low electricity price signals.

Key contributions of this chapter are: i) multi-objective energy dispatch model developed for AC/DC micro-grid, ii) maximizing renewable power penetration with the proposed micro-grid energy management (MGEM) system, iii) development of 100% renewable-based isolated micro-grid, iv) a planning model proposed to find the most appropriate configuration of energy storage system based on minimization of total life cycle cost, v) impact of demand response program included in MGEM to evaluate techno-economic benefits, vi) optimal capacity of PV+BESS system has been determined considering resiliency factor, and vii) the study carried out on grid-connected and stand-alone micro-grids considering load growth factor. The proposed planning model is solved using CPLEX solver in GAMS.

8.1.1 Techniques and Classification of Energy Storage Technologies Used in Hybrid AC/DC Micro-Grids

Due to limitations in fossil fuels and emission constraints, a large share of intermittent renewables integration to the grid is increasing. In this scenario, application of energy storage systems could help the micro-grid operator to ensure balance between the generation and load demand. Various types of energy storage systems are available as illustrated in Figure 8.2 [3].

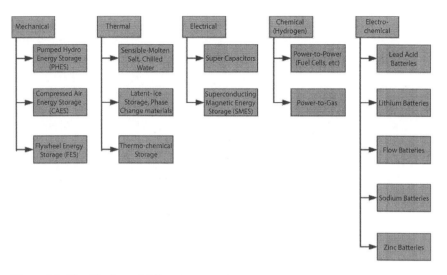

Figure 8.2 Classification of different energy storage systems.

Advanced energy storage technologies are capable of dispatching electricity within seconds and can provide power back-up ranging from minutes to many hours. The main function of energy storage system is to store energy, hold it securely and make it available instantaneously when required. Each energy storage system exhibits different characteristics of efficiency and response time, which helps to select a suitable energy storage system for a particular application. The key driving factors for rapid growth of energy storage market are: RES integration, system flexibility, ancillary services, power quality requirement, network reliability, peak load management, electric vehicles adoption, etc.

8.1.2 Applications and Benefits of Energy Storage Systems in the Microgrid System

The slogan of "Power to All for 24/7" has the additional requirement of flexible power generation. Integration of renewable-based generation is increasing rapidly due to the strong commitment of many countries to develop the clean energy power sector and ensure energy security and sustainability. Maintaining balance between generation and load demand is critical in the presence of intermittent renewables. An energy storage system is an attractive choice to maintain power balance and achieve many benefits at grid, transmission, distribution and market levels. The majority of deployment of energy storage systems at grid scale will be driven by RES integration, Ancillary Service market and T&D deferral. Further, the development of Electric Vehicle (EV) in the transportation sector and its ambitious targets increase the demand for energy storage systems. Combination of RES and energy storage system is an appropriate solution to provide flexible power supply. Further, energy storage system coupled with PV is an economic solution to provide electricity to isolated areas.

Deployment of BESS in power networks has been increasing over the last decade. Major applications of BESS in the power networks are to charge the batteries during lower electricity rate, or surplus energy available from RES and discharging back to loads during high electricity rates and to ensure balance between generation and load demand. An energy storage system offers significant benefits at each level of the power system from generation to end user. Large-scale BESS at utility level can be used to store surplus power generated during off-peak periods and delivered to maintain power balance at peak periods. In addition to this, emergency back-up power for critical loads (industries, workshops, office complexes, etc.) and other grid support features to maintain stability are other applications of small-scale BESS. Energy storage systems are also being used for critical applications like process industries and database centers to provide reliable and quality power supply. IEEE guide 2030.2.1 [18] provides recommended approaches and practices for design, operation, maintenance and integration of battery energy storage systems in power systems.

8.1.2.1 Applications and Benefits of BESS in Micro-Grid

Different applications are available with BESS at different layers of a power system from generation, transmission and distribution networks [2, 5, 18]. A summary of these major applications and associated benefits is given in Figure 8.3. The energy storage system plays an important role in efficient use of intermittent RES. Combining RES along with energy storage system helps remove this uncertainty and enables the provision of reliable power supply. At very high penetrations of variable wind and solar generation, energy storage can be effective for storing excess energy at certain times and moving it to other times, enhancing reliability and providing both economic and environmental benefits.

8.1.2.1.1 Renewable Energy Sources Integration
The variability and intermittent nature of RES cause instability of the grid. The distribution generation offers additional opportunities for small-scale

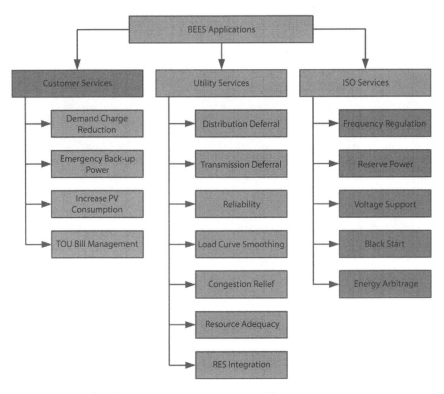

Figure 8.3 Benefits of battery energy storage systems [4].

to large-scale energy storage systems. BESS is an effective tool to overcome the above problems; it enables optimal integration of renewable energy sources, and ensures grid balancing and reliability. Moreover, significant economic benefits can also be obtained by incorporating BESS along with high-scale solar and wind farms. For small-scale systems like solar on a residential rooftop, substation and other buildings rooftop applications, using BESS electricity is stored during high PV production and dispatched to cater to load demand in an economic way during non-production time.

8.1.2.1.2 System Reliability

The main objective of a network operator is to provide 24/7 uninterrupted and quality power supply to all consumers. In addition to this, due to environmental constraints, share of clean energy sources is increasing rapidly. However, maintaining power supply reliability is a major concern with uncertainty of renewable energy sources. In this context, optimal incorporation of BESS is capable of maintaining energy balance between generation and demand. In this way reliability of small-scale to large-scale power networks can be ensured using optimal installation of BESS.

8.1.2.1.3 Voltage Control

Integration of BESS can be used to control over voltages occuring during high power production from renewables. Also, significant voltage drop observed during large motor start-up and high loading conditions. Bus voltage can be maintained within a permissible range through reactive power control using smart inverters of BESS.

8.1.2.1.4 Peak Load Shaving

Shaving peak load demand is the most important application of BESS. BESS is charged during low power consumption, lower electricity price and excess renewable power generation. The energy stored in BESS is dispatched to meet peak load demands to relieve congestion in the network. The energy demand and peak load demand is growing rapidly and integration of renewables is also increasing accordingly.

8.1.2.1.5 Frequency Response

The grid is experiencing frequency fluctuations due to large-scale penetration of the intermittent nature of renewable energy sources. Incorporation of adequate capacity of BESS provides instantaneous energy dispatch, and therefore BESS is a better option for frequency regulation in the ancillary service market. To facilitate this service, appropriate capacity and technology of BESS shall be considered with SOC factor.

8.1.2.1.6 Emergency Back-Up/Black Start

When the normal power system fails to provide stable voltage due to faults, the BESS can serve as the main power source. The BESS operate in black start mode so that voltage can be gradually restored from zero to the rated value. In the emergency back-up power case, when the normal power supply fails, BESS provide instantaneous power to ensure reliability.

8.1.3 Importance of Appropriate Configuration of Energy Storage System in Micro-Grid

Optimal operation of BESS is the most important to achieve the benefits specified in section 8.1.2. Therefore, appropriate control scheme shall be selected for effective management of BESS. Popular control methods of BESS are centralized, decentralized and coordinated approaches as described in sections 8.1.3.1-8.1.3.3. Further, topologies of BESS and power conversion system (PCS) are demonstrated in section 8.1.3.4 [18].

8.1.3.1 Decentralized Control

As shown in Figure 8.4, charge/discharge pattern, the state of charge of the battery shall be controlled by respective local controller in a decentralized approach. Therefore, minimal communication infrastructure is required in this scheme. One downside to the use of the decentralized method is that the structure of it, having no communication between the network buses, does not permit the support of other buses by neighboring buses. A bus that is distressed by overvoltage, as a result of the co-located BESS being fully charged, or due to insufficient power capacity to deal with a situation, cannot get support from other BESS in the decentralized control scheme. Decentralized control scheme adopted in [6], for frequency regulation using multiple energy storage systems.

8.1.3.2 Centralized Control

Control scheme of centralized approach is shown in Figure 8.5. The central controller continuously monitors energy balance and voltage profile throughout the network. In the centralized control method, the set-points, charge and discharge pattern of BESS are determined and set by a central controller. Therefore, strong bidirectional communication link is required in this scheme. As a result of the requirement for communication infrastructure, the centralized control method is expensive compared to decentralized control scheme. In [7], a centralized coordination scheme is proposed for small control of energy storage and distributed generation in micro-grids.

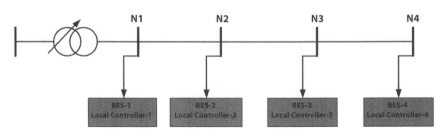

Figure 8.4 Decentralized control scheme.

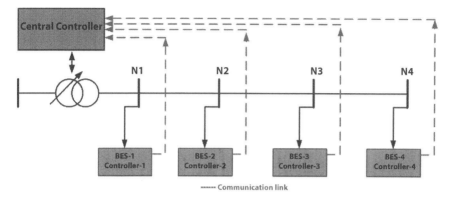

Figure 8.5 Centralized control scheme.

8.1.3.3 Coordinated Control

The coordinated control [8, 9] scheme, as shown in Figure 8.6, is a combination of all benefits of distributed control and the centralized control to deliver the best results. In [10], a control algorithm was proposed for the coordination of BESS used in low voltage power distribution system having rooftop PV systems. BESS can also be used to mitigate occurrence of overvoltage in the presence of high PV power output. Consensus control was used in their work to share the action responsibilities among the participating BESS on the network. It takes into account the system of cycling the batteries in a way to protect battery health.

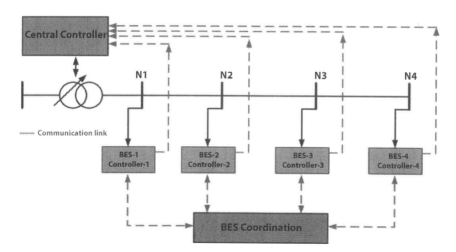

Figure 8.6 Hybrid control scheme.

8.1.3.4 Topology of BESS and PCS

Typical connection of BESS and monitoring, information exchange and control center of BESS to SCADA is shown in Figure 8.7.

Popular topologies of BESS and PCS are shown in Figure 8.8 [18]. The simplest single-stage conversion topology is shown in Figure 8.8 (a). The main drawback of this system is lack of flexibility and reliability. To enhance flexibility and reliability of the system, multiple BESS and AC/DC converters are used as shown in Figure 8.8 (b). If any BESS or AC/DC converter fails, the rest of the BESS sub-system can still work. Two-stage conversion topology is shown in Figure 8.8 (c), in which DC/DC and DC/AC converters are used. This PCS topology has strong adaptability and achieves more flexible capacity configuration due to the dc/dc link. For ease of capacity expansion, multiple DC/DC and DC/AC converters are used as shown in Figure 8.8 (d) and Figure 8.8 (e).

8.1.3.5 Battery Management System

The intent of a battery management system (BMS) [11, 18] is to monitor the healthiness of battery operation status including operational parameters of voltages, currents, battery internal and ambient temperature, SOC, SOH, etc., and alarm indications sent in case of abnormality. BMS is essential

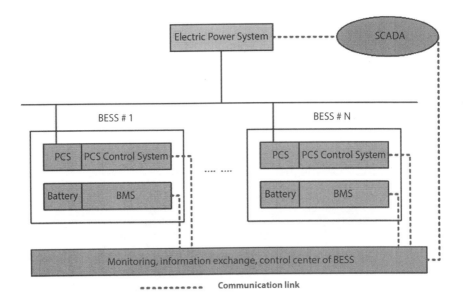

Figure 8.7 Block diagram of connection between BESS and SCADA [18].

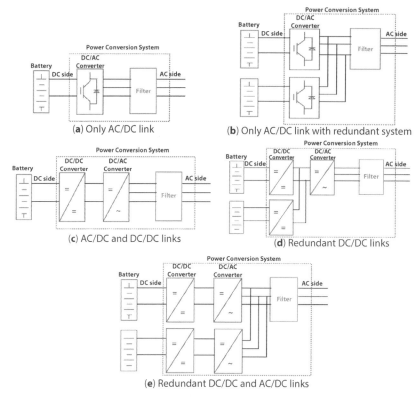

Figure 8.8 Battery and power conversion topologies [18].

for optimal utilization of energy inside the battery and maintaining its life time in line with design. Key features of BMS are shown in Figure 8.9.

8.2 Concept of Micro-Grid Energy Management

Expansion of generation, transmission and distribution systems has been increasing to meet growing electricity demand. However, today approximately 13% of world population has no access to electricity [12]. The present electricity supply systems are unsustainable—environmentally, economically and socially. There is a need for rapid transformation from conventional energy system to a low-carbon, efficient and environmentally friendly energy system. The distribution networks are active in nature with deployment of distributed generation. Due to limitation of availability of fossil fuels and high fuel prices, governments around the world have

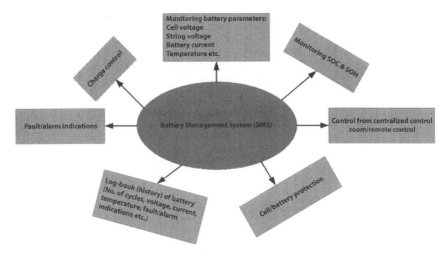

Figure 8.9 Features of battery management system.

positive attitudes towards development of the clean energy power sector with renewables.

8.2.1 Concept of Micro-Grid

Micro-grid refers to a local grid having distributed energy resources and loads that can operate in a controlled and coordinated manner in grid-connected and stand-alone modes [2]. The technical details of distributed energy resources have been specified in IEEE Std 1547 [19], "IEEE Standard for Interconnecting Distributed Resources with Electric Power Systems". Micro-grids are resilient because they have their own power generation and operate uninterruptedly during storms and other major power outages. Further, the major benefits of micro-grids are: increased power reliability, reduced utility costs and improved economic competitiveness, reduced greenhouse gas emissions, lowered transmission and distribution losses and capable of operating on renewable or non-renewable resources.

A micro-grid system with multiple energy sources shall be carefully designed with appropriate mix of energy sources and energy storage systems for efficient and economic operation of the system. A typical micro-grid system with hybrid energy sources and storage system is shown in Figure 8.10. Unprecedented generation – demand events shall be considered to demonstrate the system's robustness and RES integration.

Responsibility of energy management in each MG units is assigned to respective MG agents as shown in Figure 8.11. MG agents are coordinated

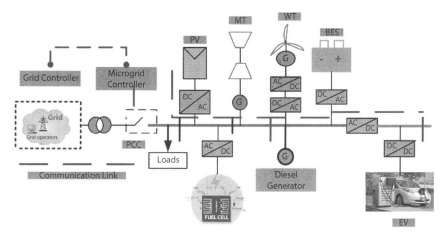

Figure 8.10 Typical representation of micro-grid.

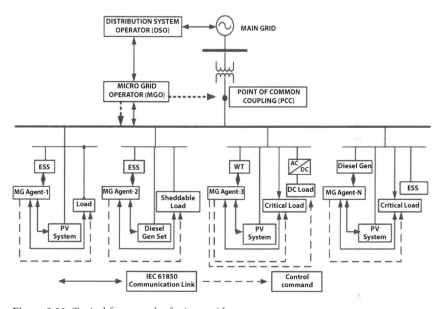

Figure 8.11 Typical framework of micro-grid.

with micro-grid operator (MGO) through bidirectional communication link for effective energy dispatch. Uncertainties of power demand and power generation shall be considered in energy dispatch problem. With adequate BESS, the micro-grid network becomes strong and stable grid. Appropriate battery technology selection and sizing of BES for micro-grid expansion was solved using mixed-integer linear programming [5].

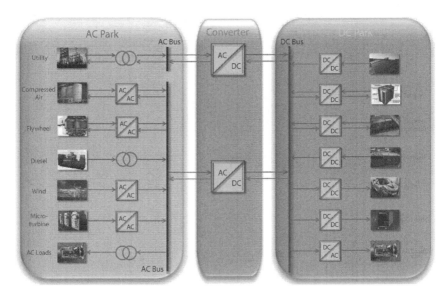

Figure 8.12 Schematic of micro-grid with hybrid energy sources [1].

Figure 8.12 shows a typical structure of a hybrid AC/DC micro-grid system [1].

8.2.2 Benefits of Micro-Grids

As mentioned, micro-grids provide an effective way for integrating small-scale distributed energy resources in proximity of load into low-voltage distribution network. Micro-grids can supply highly reliable power to a wide range of customers, both residential and commercial, such as schools, hospitals, warehouses, shopping centers, university campuses, military installations, data centers, etc. It is also useful for remote places having no or limited access to the utility grid. Further, micro-grid technology can also be used in areas facing high stress and congestion in their transmission and distribution systems. Following are the major benefits of micro-grids: facilitate integration of wind- and solar-based generation, curb fossil fuel dependency, carbon emissions and promote energy sustainability, increase power quality and reliable power supply to end users. Controllable loads in the micro-grid can participate in demand response, which can contribute to peak shaving during times of peak demand by reducing their own consumption via shedding of non-critical loads and delivering more power to the main grid utility. Micro-grids can lower overall distribution system losses and voltage management by implementing distributed generation located at the demand site, eliminating the need for transmission lines and deferring the construction of new transmission lines to a later time. Therefore, considerable revenue can be generated to MG network operator with RES and BESS [1, 2].

8.2.3 Overview of MGEM

As illustrated in Figure 8.13, micro-grid energy management (MGEM) involve following main blocks of monitoring (forecast load demand, renewable power generation, utility electricity prices, etc.) controlling (Distributed Energy Resources ON/OFF control, switching of controllable loads, battery SOC, power import/export from grid) and optimization to achieve minimum cost of energy, maintain supply-demand balance and reliable power supply. Energy management module of the central controller is responsible for optimal energy dispatch in MG. The problem of MGEM involves finding the optimal unit commitment (UC) and dispatch of the controllable DGs such that the certain objectives are achieved. The topology and operation of the micro-grid is more complex than the traditional power distribution system due to the presence of multiple energy sources, storage devices, integration with main grid and loads in a single entity. Further, energy balance is a more challenging issue in micro-grids operating in stand-alone mode due to availability of limited energy sources in presence of uncertainty of renewable power generation. Therefore, it is noteworthy that implementation of MGEM system is essential for optimal operation of micro-grid with effective utilization of energy sources for grid resiliency, energy balance, rural electrification, reducing the cost of

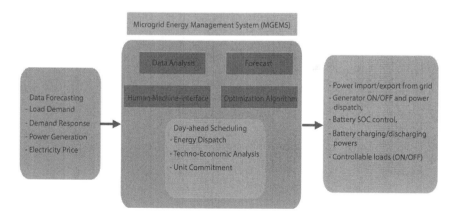

Figure 8.13 Micro-grid energy management system.

electricity and reducing carbon emissions. For effective micro-grid energy management, bidirectional communication link is essential between micro-grid controller and local controllers, and between grid controller and micro-grid controller.

8.3 Modelling of Renewable Energy Sources and Battery Storage System

A micro-grid system comprising hybrid renewable sources and energy storage system could be an economic solution to produce reliable clean energy. Modelling of each component is explained in the subsections.

i. PV System
PV power output is determined using equations (8.1), (8.2).

$$P_{pv} = P_{pv}^r f_{pv} \left(\frac{\overline{G_T}}{G_{T,STC}} \right) \left[1 + \alpha_p \left(T_c - T_{c,STC} \right) \right] \qquad (8.1)$$

$$T_c = T_{a+} \left(T_{c,NOCT} - T_{a,NOCT} \right) \left(\frac{G_T}{G_{T,NOCT}} \right) \left(1 - \frac{\eta_{mp}}{0.9} \right) \qquad (8.2)$$

ii. Wind Power
Wind turbine power output is computed using equations (8.3), (8.4).

$$P_w = P_r \begin{cases} 0, v \leq v_{ci} \\ P_n(v), v_{ci} < v < v_r \\ 1, v_r < v < v_{co} \\ 0, v > v_{co} \end{cases} \qquad (8.3)$$

$$P_{wt} = \eta_{wt} P_w \qquad (8.4)$$

iii. Battery Energy Storage System
Capacity of a battery is affected by battery configuration, back-up period, temperature, battery life time, depth of discharge and renewable energy sources, etc. Detailed modeling of BESS is presented considering

Role of Energy Storage Systems in the Micro-Grid Operation

charge/discharge limits, depth of discharge (DOD), efficiency and battery life time. Power charging/discharging of battery is expressed in equation (8.5).

$$P_{BES}(t) = \begin{cases} P_{ch}(t) & \text{if } P_{PV}(t) + P_{WT}(t) + P_{DG}(t) + P_g(t) - P_d(t) > 0 \\ P_{dch}(t) & \text{if } P_{PV}(t) + P_{WT}(t) + P_{DG}(t) + P_g(t) - P_d(t) < 0 \end{cases} \quad (8.5)$$

Charging and discharging energy of battery is calculated using equation (8.6) and equation (8.8).

Charging mode:

$$E_{ch}(t) = \left(\frac{P_{DG}(t) + P_{WT}(t) + P_{FC}(t) + P_{MT}(t) - P_d(t)}{\eta_{Conv}} + P_{pv}(t) \right) * \Delta t * \eta_{ch} \quad (8.6)$$

$$SOC(t) = SOC(t-1)(1-\sigma) + E_{ch}(t) \quad (8.7)$$

Discharging mode:

$$E_{dch}(t) = \left(\frac{-P_{DG}(t) - P_{WT}(t) - P_{FC}(t) - P_{MT}(t) + P_d(t)}{\eta_{Conv}} - P_{pv}(t) \right) * \Delta t * \eta_{dch} \quad (8.8)$$

$$SOC(t) = SOC(t-1)(1-\sigma) + E_{ch}(t) \quad (8.9)$$

State of charge (SOC) of battery is modeled using equation (8.7) and equation (8.9). Depth of discharge (DOD) of a battery indicates the capacity of discharged as a fraction of the initial capacity as expressed in equation (8.10).

$$DOD = \frac{\int_{t0}^{tb} I_{dch}.dt}{CBES_0} \quad (8.10)$$

where $CBES_0$ is the initial battery capacity, tb is the time at the end of the evaluation period, is the initial time and I_{dch} is the current from the battery.

State of health (SOH) of battery, indicates battery capacity as a fraction of nominal battery capacity declared by manufacturer as expressed in equation (8.11).

$$SOH(t) = \frac{CBES_{ref}(t)}{CBES_{nom}} \quad (8.11)$$

where $CBES_{nom}$ is the nominal capacity of reference available by manufacturer and $CBES_{ref}(t)$ is the capacity at the time of estimating the SOH. Change in battery energy is expressed in equation (8.12).

$$\Delta E_{BES} = E_{BES}(t) - E_{BES}(t-1) \quad (8.12)$$

Two independent factors may limit the lifetime of the storage bank: the lifetime throughput ($Q_{lifetime}$) and the storage float life ($R_{batt,f}$). The storage bank life and battery wear cost is determined using equations (8.13) and (8.14).

$$R_{batt} = \begin{cases} \dfrac{N_{batt} Q_{lifetime}}{Q_{thrpt}} & \text{if limited by throughput} \\ R_{batt,f} & \text{if limited by time} \\ \min\left[\dfrac{N_{batt} Q_{lifetime}}{Q_{thrpt}}, R_{batt,f}\right] & \text{if limited by throughput and time} \end{cases} \quad (8.13)$$

$$C_{bw} = \frac{C_{rep,batt}}{N_{batt} Q_{lifetime} \sqrt{\eta rt}} \quad (8.14)$$

iv. Power Converter
A converter is required in hybrid systems that contain AC and DC elements. Rating of inverter is determined using equation (8.15).

$$INV_{cap} = (3L_{ind}) + L_0 \quad (8.15)$$

v. Generator Capacity
The output power of controllable DG unit shall be within its upper and lower limits as follows.

$$P_{DG}^{min} \leq P_{DG}(t) \leq P_{DG}^{max} \quad (8.16)$$

where, P_i^{min} and P_i^{max} represent minimum and maximum generation capacity of ith unit.

Emission cost of diesel generator is as follows:

$$EC_{DG}^t = k_c C_e P_{DG}^t \qquad (8.17)$$

vi. Demand Response
Micro-grid operator offers incentive to consumers against participation in demand response program. Incentive cost for demand response is:

$$IC_t^{DR} = \sum_{b \in nb} k_{DR} P_{b,t}^{DR} \qquad (8.18)$$

vii. Loss of Power Supply Probability
Loss of power supply probability (LPSP) is a design indicator which measures probability of unmet energy. LPSP is the probability of deficit power to supply load demand. LPSP is the ratio of unmet energy to total energy demand as given in equation (8.19).

$$LPSP = \frac{\sum_{t=1}^{T} D_{unmet,t}}{\sum_{t=1}^{T} D_t} \qquad (8.19)$$

$$\text{Availability of power supply} = (1 - LPSP) \qquad (8.20)$$

viii. MGEM Problem Modelling
MGEM problem is formulated as mixed integer linear programming problem and implemented in GAMS 23.4 environment and solved using CPLEX solver. MGEM problem involves finding unit commitment, power exchange with main grid, output power of controllable DGs and input/output power of BESS.

a. Objective Function
Objective function includes minimization of total cost as defined in equation (8.21) subject to satisfying the constraints.

$$\min\{F_1(P_g), F_2(P_{DG}), F_3(C_{RES,i}(P_{RES,i}(t))), F_4(CE), F_5(DR), F_6(P_{loss})\} \qquad (8.21)$$

$$F_1(P_g) = \sum_{t=1}^{n} \{C_g(t)P_g(t)\} \tag{8.22}$$

$$F_2(P_i) = \sum_{t=1}^{n} \left\{ \sum_{i=1}^{NDG} FC_i(P_i(t)) + S_i(t) \right\} \tag{8.23}$$

$$F_3(C_{RES,i}(P_{RES,i}(t))) = (a_{RES,i} P_{RES,i}(t)^2 + b_{RES,i} P_{RES,i}(t) + C_{RES,i}) \tag{8.24}$$

$$F_4(CE_i) = \sum_{t=1}^{n} \left\{ \sum_{i=1}^{N} \sum_{j=1}^{M} (EF_{ij}.P_i(t))ce_{dg} + \sum_{j=1}^{M} (EF_{gj}.P_g(t))ce_g \right\} \tag{8.25}$$

$$F_5(DR) = IC_t^{DR} \tag{8.26}$$

$$F_6(P_{loss}) = K_e TPL \tag{8.27}$$

$$FC_i(P_i(t)) = (a_i P_i(t)^2 + b_i P_i(t) + C_i) \tag{8.28}$$

$$S_i(t) = SC_i \text{ if } \theta_i(t) - \theta_i(t-1) = 1 \tag{8.29}$$

b. Constraints
Following constraints are considered in MGEM problem.

- Power Balance Constraint

$$P_D(t) + P^{DR}(t) + P_{loss}(t) + P_{ch}(t) = P_g(t) + P_{DG}(t) + P_{WT}(t) + P_{PV}(t) + P_{dc}(t) \tag{8.30}$$

- Generation Capacity Constraint

$$P_i^{min} \leq P_i(t) \leq P_i^{max} \tag{8.31}$$

- Generation Constraint of RES

$$P_{pv,i} \leq P_{pv}^{max} \quad (8.32)$$

$$P_{wt,i} \leq P_{wt}^{max} \quad (8.33)$$

- Consumer Loads

Based on process/operation requirements loads are categorized as critical loads, non-critical loads, transferrable, sheddable and non-sheddable loads, etc.

$$0 \leq P_{L,t}^{trans} \leq P_{L,t}^{trans,max} \quad (8.34)$$

$$0 \leq P_{b,t}^{DR} \leq \propto P_{D\ b,t} \quad (8.35)$$

- Charging-Discharging Constraints

$$0 \leq P_{ch}(t) \leq P_{BES}^{r} \quad (8.36)$$

$$0 \leq P_{dch}(t) \leq P_{BES}^{r} \quad (8.37)$$

The output power of each energy storage unit must satisfy charge-discharge limits as follows.

$$ES_i^{min} \leq ES_i(t) \leq ES_i^{max} \quad (8.38)$$

where, $ES_i^{min}n$ and ES_i^{max} represent the minimum and maximum exchanged power of energy storage unit i, respectively.

$ES_i(t) > 0$ *energy storage unit is in discharging mode*
$ES_i(t) < 0$ *energy storage unit is in charging mode*

$$SOC_i(t+1) = SOC_i(t) - \frac{\eta_i ES_i(t)}{C_i} \quad (8.39)$$

$$SOC_i^{min} \leq SOC_i(t) \leq SOC_i^{max} \quad (8.40)$$

- Reserve Power

Grid operators must have planned for adequate amount of reserve power capacity at strategic locations in the grid to ensure reliable power supply.

$$P_{res,AC} = r_{load} * P_{prime,AC} + r_{peakload} * \overline{P_{prime,AC}} + r_{wind} * P_{windAC} \quad (8.41)$$

$$P_{res,DC} = r_{load} * P_{prime,DC} + r_{peak\ load} * \overline{P_{prime,DC}} + r_{solar} * P_{PV} \quad (8.42)$$

8.4 Uncertainty of Load Demand and Renewable Energy Sources

The uncertainty of electricity demand is determined using normal distribution [13, 14] as given in equation (8.43) and equation (8.44).

$$f(P_{L,i}) = \left(\frac{1}{\sigma_{PL,i}\sqrt{2\pi}}\right) exp^{-\frac{(P_{L,i}-\mu_{PL,i})^2}{2\sigma_{PL,i}^2}} \quad (8.43)$$

$$f(Q_{L,i}) = \left(\frac{1}{\sigma_{QL,i}\sqrt{2\pi}}\right) exp^{-\frac{(Q_{L,i}-\mu_{QL,i})^2}{2\sigma_{QL,i}^2}} \quad (8.44)$$

The load growth factor is modeled in equations (8.45) and (8.46) for future planning and expansion of the distribution systems.

$$PD_i = P_o(1+g)^T \quad (8.45)$$

$$QD_i = Q_o(1+g)^T \quad (8.46)$$

where, P_0, Q_0 = initial real and reactive power load demand, PD_i, QD_i = real and reactive power load demand considering load growth, g = annual growth rate, T = planning period up to which the feeder can take the load and g = 7% and T = 5 years.

Renewable Uncertainty Modelling

In this work uncertainty of PV and wind turbine power output is modeled using beta distribution function [14] and weibull distribution function [13], respectively.

$$f_b(s) = \frac{s^{\alpha-1}(1-s)^{\beta-1}}{\Gamma(\alpha).\Gamma(\beta)}\Gamma(\alpha+\beta) \quad (8.47)$$

$$\alpha = \mu(\frac{(1-\mu)\mu}{\sigma} - 1) \qquad (8.48)$$

$$\beta = (1-\mu)(\frac{(1-\mu)\mu}{\sigma} - 1) \qquad (8.49)$$

$$P_{PV} = \begin{cases} P_{PV\_rated} * \dfrac{s}{s_r}, & 0 < s < s_r \\ P_{PV\_rated} s_r < s \end{cases} \qquad (8.50)$$

$$f(v) = \frac{k}{c}(\frac{v}{c})^{k-1} \exp(\frac{-v}{c})^k, \quad 0 \le v \le \infty \qquad (8.51)$$

$$k = (\frac{\sigma}{\bar{v}})^{-1.086} \qquad (8.52)$$

$$c = \frac{\bar{v}}{\Gamma(1+\dfrac{1}{k})} \qquad (8.53)$$

8.5 Demand Response Programs in Micro-Grid System

Demand-side participation is an important aspect for optimal energy scheduling at lower cost and higher energy security [15, 16]. Demand response (DR) is one of the most popular methods of demand-side management that encourages customers to adjust their elastic loads in accordance with the operator's request or price signals. Major benefits of a demand response program are: cost saving, optimal operation, reducing the use of costly generators, reduced purchase of expensive power from the main grid and load curve flattening. DR programs are classified into two main categories: time-based rate (TBR) and incentive-based (IB) program.

8.5.1 Modelling of Price Elasticity of Demand

Elasticity is defined as the load sensitivity with respect to the electricity price as expressed in equation (8.54).

$$E = \frac{\Delta d/d_0}{\Delta p/p_0} \tag{8.54}$$

where Δd and Δp are the change in demand and price respectively and d_0 and p_0 are the initial demand and price, respectively.

Elasticity is composed of two different coefficients namely self-elasticity $E(i,i)$ and cross elasticity $E(i,j)$.

$$E(i,i) = \frac{\partial d(t_i)/d_0}{\partial p(t_i)/p_0}, \leq 0 \tag{8.55}$$

$$E(i,j) = \frac{\partial d(t_i)/d_0}{\partial p(t_i)/p_0}, \geq 0 \tag{8.56}$$

Price elasticity matrix will be of the order 24x24 for 24-h of a day as represented in equation (8.57).

$$\begin{bmatrix} \frac{\Delta d(1)}{d_0(1)} \\ \vdots \\ \frac{\Delta d(24)}{d_0(24)} \end{bmatrix} = \begin{bmatrix} E(1,1) & \cdots & E(1,24) \\ \vdots & \ddots & \vdots \\ E(24,1) & \cdots & E(24,24) \end{bmatrix} \times \begin{bmatrix} \frac{\Delta p(1)}{p_0(1)} \\ \vdots \\ \frac{\Delta p(24)}{p_0(24)} \end{bmatrix} \tag{8.57}$$

The electricity prices are assumed as 0.03 $/KWh in flat rate, 0.012, 0.02 and 0.05 $/KWh at valley, off-peak and peak periods, respectively. Incentive and penalty rate assumed as 0.025 $/KWh and 0.01 $/KWh.

As specified in Table 8.1, the load curve is divided into three different periods i.e., valley, off-peak and peak period.

Table 8.1 Self and cross elasticity for 24-hour.

	Valley	Off-peak	Peak	Time period
Valley	-0.1	0.01	0.012	1-9
Off-peak	0.01	-0.1	0.016	10-18
Peak	0.012	0.016	-0.1	19-24

8.5.2 Load Control in Time-Based Rate DR Program

Modified load demand due to the implementation of time-based rate DR program is obtained from the following equation.

$$d(i) = d_o(i)\{1 + \frac{E(i)[\rho(i) - \rho_o(i)]}{\rho_o(i)}$$

$$+ \sum_{j=1, j \neq i}^{24} E(i,j) \frac{[\rho(j) - \rho_o(j)]}{\rho_o(j)}\} \, i = 1, 2, \ldots 24 \quad (8.58)$$

8.5.3 Load Control in Incentive-Based DR Program

Modified load demand due to the implementation of incentive-based DR program is obtained as follows.

$$d(i) = d_o(i)\{1 + \frac{E(i)[\rho(i) - \rho_o(i) - A(i) + \text{pen}(i)]}{\rho_o(i)}$$

$$+ \sum_{j=1, j \neq i}^{24} E(i,j) \frac{[\rho(j) - \rho_o(j) - A(j) + \text{pen}(j)]}{\rho_o(j)}\} \quad (8.59)$$

8.6 Economic Analysis of Micro-Grid System

The objective is to optimize various possible configurations and rank each feasible configuration based on net present cost (NPC). Capital cost, replacement cost, operation and maintenance cost, and fuel cost are included in NPC calculation. NPC is calculated using equation (8.60).

$$NPC = \frac{C_{ann, total}}{CRF(i, N)} \quad (8.60)$$

Capital recovery factor (CRF) is determined using equation (8.61).

$$CRF(i, N) = \frac{i(1+i)^N}{i(1+i)^N - 1} \quad (8.61)$$

$$i = \frac{i' - f}{1 + f} \quad (8.62)$$

$$\text{Levelized COE} = \frac{C_{ann,\,total}}{E_{served}} \qquad (8.63)$$

$$NPC = C_{PV} + C_{WT} + C_{DG} + C_{BES} + C_{CONV} \qquad (8.64)$$

C_{PV}, C_{WT}, C_{DG}, C_{BES} and C_{CONV} are the sum of present value of capital cost, O&M cost, replacement cost and fuel cost of PV, WT, DG, BES and power converter, respectively.

8.7 Results and Discussions

In this chapter, grid-connected and stand-alone micro-grid systems are simulated with hybrid energy sources and energy storage devices. The role of BESS has been studied for various applications in micro-grid. Simulation results are obtained for optimal capacity of PV, WT, diesel generator, converter, BESS and state of charge of battery. Real-time data of solar irradiance and wind velocity is taken from reference [17]. The hybrid system is designed to provide requisite operating reserve to ensure grid reliability. A comparative analysis of techno-economic benefits is obtained for different configurations.

8.7.1 Dispatch Schedule Without Demand Response

Different micro-grid systems are simulated with PV/WT/BES systems considering hourly variation of load demand, solar radiation, wind velocity and ambient temperature. Uncertainty of wind speed and solar irradiance is shown in Figure 8.14. Renewable-based hybrid energy systems play a key role in developing the green power sector and lowering the cost of energy.

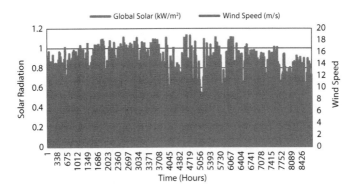

Figure 8.14 Hourly variation of solar radiation and wind speed.

Table 8.2 Annual energy production of for various configurations.

Configuration	Resource	Annual energy production (kWh/year)	Fraction (%)
PV+WT+Grid+BES	PV	83365	20.1
	WT	34504	8.31
	Grid	297555	71.6
PV+Grid+BES	PV	154398	38.3
	Grid	248378	61.7
WT+Grid+BES	WT	51756	12.5
	Grid	361913	87.5
Grid	Grid	406427	100
DG	DG	406427	100

A comparative analysis of techno-economic benefits is given in Tables 8.2–8.4 for various configurations. Annual energy production details of each configuration are given in Table 8.2 for annual energy consumption of 406427 kWh/year which is being met by PV power production of 83365 kWh/year, grid power purchase of 297555 kWh/year and wind turbine power production of 34504 kWh/year in PV+WT+Grid+BES system. Lower COE observed in PV+Grid+BES system compared to other configurations is specified in Table 8.3. Capacity of each micro-grid system is given in Table 8.3. Annual NPC is low for PV+Grid+BES system (616420$) and high for micro-grid with WT+Grid+BES system (937146.5$). Hourly optimal power dispatch and SOC of battery of hybrid system is illustrated in Figure 8.15.

Reduction in greenhouse gas emissions is given in Table 8.4 for renewable based micro-grids compared to conventional power system. The hybrid system PV+WT+Grid+BES reduces CO_2 emissions by 48.48% per year as compared to micro-grid system operating with diesel generator.

8.7.2 Dispatch Schedule with Demand Response

Impact of demand response on energy dispatch and techno-economic benefits in micro-grid with RES and storage devices has been investigated. Variation in load demand in presence of demand response program is shown in Figure 8.16. It has been observed from the simulation results that demand response

Table 8.3 Simulation results of hybrid micro-grid system.

	Configuration	PV (kW)	WT (10kW)	BES (strings)	Grid (kW)	Converter (kW)	NPC ($)	COE	Operating cost ($)
Without load growth	PV+WT+BES	50	6	309	50	48	843439.6	0.164	51358.64
	PV+BES	80	---	145	50	54	616420	0.125	37905
	WT+BES	---	9	338	50	43.2	937146.5	0.1823	59998.59
With load growth	PV+WT+BES	89.4	12	753	50	102	1529342	0.213	87905.49
	PV+BES	110	---	456	50	61.4	1220994	0.179	75397.22
	WT+BES	---	20	772	60	129	1584739	0.2197	93446.21

Table 8.4 Greenhouse gas emissions summary.

Emissions (kg/yr)	Without load growth					With load growth				
	DG	Grid	PV+grid+BES	WT+grid+BES	PV+WT+grid+BES	DG	Grid	PV+grid+BES	WT+Grid+BES	PV+WT+grid+BES
CO_2	304701	256862	156975	228729	188,055	572729	360263	227616	298580	236343
CO	1439	0	0	0	0	403	0	0	0	0
Unburned Hydrocarbons	83.6	0	0	0	0	30.3	0	0	0	0
Particulate Matter	8.24	0	0	0	0	34.7	0	0	0	0
SO_2	744	1114	681	992	815	1422	1562	987	1294	1025
NO_x	164	545	333	485	399	5765	764	483	633	501

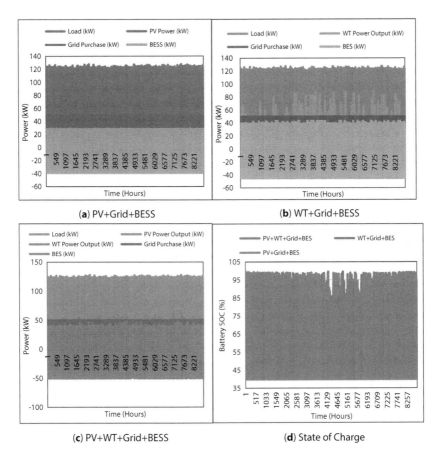

Figure 8.15 Hourly power dispatch and state of charge of BESS.

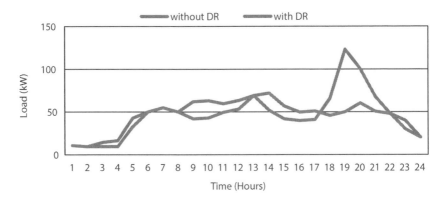

Figure 8.16 Variation in load demand during DR implementation.

Table 8.5 Annual energy production considering demand response program.

Configuration	Resource	Annual energy production (kWh/year)	Fraction (%)
PV+WT+Grid+BES	PV	83365	20.1
	WT	5751	1.39
	Grid	325148	78.5
PV+Grid+BES	PV	154398	37.2
	Grid	260688	62.8
WT+Grid+BES	WT	69008	16.9
	Grid	339146	83.1

plays an important role in economic operation of micro-grid. Summary of techno-economic benefits considering impact of demand response is given in Tables 8.5–8.7. Annual energy production details of each configuration are given in Table 8.5 for annual energy consumption of 402312 kWh/year which is being met by PV power production of 83365 kWh/year, grid power purchase of 325148 kWh/year and wind turbine power production of 5751 kWh/year in PV+WT+Grid+BES system. Table 8.6 shows the capacity of each micro-grid system. As mentioned in Table 8.6, levelized COE is low for PV+Grid+BES system (0.0888$/kWh) and high for WT+Grid+BES system (0.1313$/kWh). Annual NPC is low for PV+Grid+BES system (460021.3$) and high for micro-grid with WT+Grid+BES system (683798.6$). Hourly optimal power dispatch and state of charge of battery of hybrid micro-grid system is shown in Figure 8.17. Greenhouse gas emission in each configuration of micro-grid is given in Table 8.7.

8.7.3 Micro-Grid Resiliency

Resiliency provided by PV + BES system is fundamentally different than that provided by a diesel generator. Generators provide adequate power supply based on fuel reserve availability to meet the required load demand. However, electricity provided by PV + BES system is uncertain in nature and mainly depends on: battery SOC at the time of the power outage and solar resource availability. Therefore, uncertainty factors need to be considered while sizing PV + BES system. In this work, optimize size of PV + BES system has been determined for cost savings and enhanced resiliency

Table 8.6 Simulation results of hybrid micro-grid system considering demand response program.

	Configuration	PV (kW)	WT (10kW)	BES (strings)	Grid (kW)	Converter (kW)	NPC ($)	COE	Operating cost ($)
Without load growth	PV+WT+BES	50	1	65	50	24.3	537685.8	0.1034	35945.5
	PV+BES	80	---	5	50	54.8	460021.3	0.0888	29035.4
	WT+BES	---	12	108	50	20.4	683798.6	0.1313	45194.43
With load growth	PV+WT+BES	90	8	299	50	59.8	1067018	0.1480	65318.39
	PV+BES	110	---	183	50	92.9	928961.4	0.1316	58289.92
	WT+BES	---	20	560	60	49.2	1310477	0.1803	79308.7

Table 8.7 Greenhouse gas emissions summary considering demand response program.

Emissions (kg/yr)	Without load growth			With load growth		
	PV+grid+BES	WT+grid+BES	PV+WT+grid+BES	PV+grid+BES	WT+grid+BES	PV+WT+grid+BES
CO_2	164755	214341	205494	225006	295483	242608
CO	0	0	0	0	0	0
Unburned Hydrocarbons	0	0	0	0	0	0
Particulate Matter	0	0	0	0	0	0
SO_2	714	929	891	976	1281	1052
NO_x	349	454	436	477	626	514

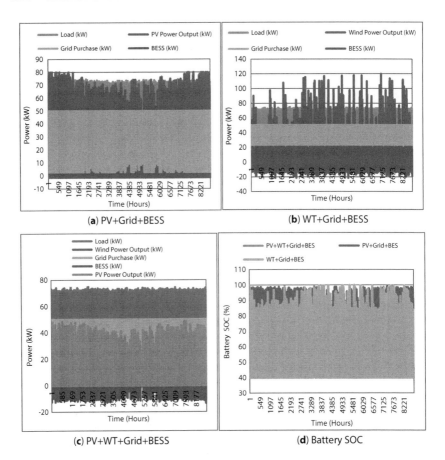

Figure 8.17 Hourly power dispatch and state of charge of BESS considering demand response.

of the system. In this case study, system is optimized to minimize lifecycle cost of energy without considering resiliency factor and then re-optimized the system taking care of resiliency.

Simulation results are presented with the cost-effective hybrid system combination of PV and BESS to sustain critical load and ensure grid resiliency. Hourly power dispatch considering resiliency is illustrated in Figure 8.18. The hybrid system includes PV system rated 449 kW and 136 kW batteries (746 kWh). This system sustains critical load of 50% for all potential 48-hour outages throughout the year. Detailed simulation results for resiliency are demonstrated in Table 8.8 including comparison of other systems. Figure 8.19 shows the amount of time that the system will sustain the critical load for outages. Outage simulation has been carried out to evaluate

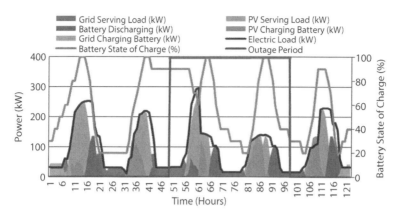

Figure 8.18 Hourly power dispatch considering resiliency.

Table 8.8 Comparison of simulation results considering resilience.

	Business as usual	Resilience	Financial
Survives Specified Outage	No	Yes	No
Average (hrs)	0	1020	10
Minimum (hrs)	0	2	0
Maximum (hrs)	0	4052	63
	System Size, Energy Production, and System Cost		
PV Size (kW)	0	449	366
Annualized PV Energy Production (kWh)	0	716554	584728
Battery Power (kW)	0	136	75
Battery Capacity (kWh)	0	746	231
Net CAPEX + Replacement + O&M ($)	0	1064278	527051
Energy Supplied From Grid in 1st Year (kWh)	992952	318458	443892
	1stYear Utility Cost - Before Tax		
Energy Cost ($)	74050	20817	31145

(*Continued*)

Table 8.8 Comparison of simulation results considering resilience. (*Continued*)

	Business as usual	Resilience	Financial
Demand Cost ($)	79692	26521	46469
Fixed Cost ($)	5551	5551	5551
	Life Cycle Utility Cost - After Tax		
Energy Cost ($)	709556	199472	298434
Demand Cost ($)	763618	254129	445269
Fixed Cost ($)	53191	53191	53191
	Business as usual	**Resilience**	**Financial**
	Total System and Life Cycle Utility Cost - After Tax		
Initial cost (before incentives) ($)	N/A	1145520	745458
Initial cost (after incentives) ($)	N/A	720625	444183
O&M and replacement costs ($)	N/A	205124	76873
Total Life Cycle Costs ($)	1526366	1365948	1323944
Net Present Value ($)	0	160418	211011

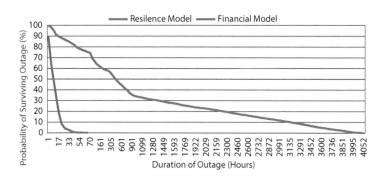

Figure 8.19 System survival for grid outages.

the amount of time that the system can survive during grid outages. It has been observed from the simulation results that capacity of PV system is increased by 22% when resiliency is considered.

8.7.4 BESS for Emergency DG Replacement

With the growing concern about rising cost of power from diesel generating sets (DG sets) as well as associated air pollution, industrial and commercial customers depending on DG sets for reliability can switch over to renewable with battery energy storage system. Moreover, many islands which are presently operating on DG sets are also switching over to hybrid power system comprised of renewable energy sources and storage system. Downward price trend of PV modules and storage system is accelerating utilities and customers in decision making of replacement of DG sets by renewable with storage system.

Table 8.9 provides a summary of simulation results for isolated micro-grid located in Andaman Islands operating with DG sets and RES + BES. Annual energy consumption of the test system is 143194 kWh/year which is being met through diesel generator rated 15 kW. To reduce cost of energy and greenhouse gas emissions, a hybrid power system is proposed with RES + BES. In the proposed hybrid system PV+WT+BES, the load demand is catered through PV power production of 55489 kWh/year and WT power production of 87705 kWh/year. It is evident from the simulation results that the project location has adequate renewable potential and unmet energy is zero throughout the planning period. It is recommended that micro-grid network operators carry out detailed techno-economic analysis and accordingly, replace existing diesel generator sets by renewable-based distributed generation coupled with battery storage system. Integration of diesel generator has stronger influence on COE than battery energy storage system. From Table 8.9, it can be observed that levelized COE is low for PV+WT+BES system (0.486 $/kWh) and high for DG system (8.46 $/kWh). Annual NPC is low for PV+WT+BES system (303129.5 $) and high for micro-grid with DG (5985793.95 $). Hourly optimal power dispatch of the hybrid system is illustrated in the Figures 8.20–8.21 to

Table 8.9 Comparison of simulation results for different systems.

	COE ($/kWh)	NPC ($)	Operating Cost ($/yr)	Fuel Cost ($)
DG	8.46	5985793.95	462911.3	5985793.95
PV+WT+BES	0.432	303129.50	5931.78	--

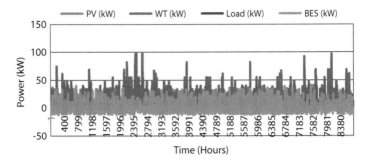

Figure 8.20 Hourly power dispatch with PV, WT and BES.

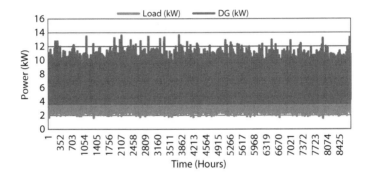

Figure 8.21 Hourly power dispatch with DG only.

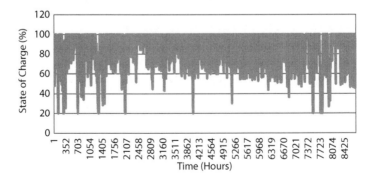

Figure 8.22 Hourly state of charge of battery.

Table 8.10 Summary of greenhouse gas emissions.

Emissions (Kg/yr)	With diesel generator	PV+WT+BES
CO_2	62987	–
CO	427	–
Unburned Hydrocarbons	17.3	–
Particulate Matter	25.9	–
SO_2	127	–
NOx	485	–

balance electricity demand and power generation subject to minimization of annual NPC. SOC of battery in the hybrid system is shown in Figure 8.22. Further, greenhouse gas emission detail is given in Table 8.10 for micro-grid with diesel generator and micro-grid with 100% renewables.

8.8 Conclusions

In this chapter, classification of various energy storage systems, applications and control topologies of BESS is presented. Further, the role of BESS in micro-grids for various benefits is also demonstrated. Required flexible power generation can be achieved with appropriate inclusion of energy storage systems into the Microgrid. In this chapter, multi-objective energy management problem is solved using GAMS/CPLEX solver. Optimal energy dispatch and techno-economic analysis is also discussed with RES and BESS. It is observed that isolated power systems have higher levelized cost of energy due to high fuel cost of diesel generators. Significant benefits of demand charge reduction, ancillary services, energy arbitrage, etc., are studied with RES+BESS. In addition, impact of demand response on energy dispatch and cost of energy are also investigated in this work. Levelized COE is reduced by 40.76% after implementing demand response program. It is observed that COE is minimum in PV+BES system (0.0888$/kWh) and most economical compared to other configurations. As demonstrated in results and discussions, an energy storage system is capable of providing valuable services throughout the micro-grid. Optimal capacity of PV+BESS system has been determined considering resiliency factor. It is noted that larger capacity of BESS required ensuring grid reliability during outages, and consequently the overall cost of the system increased.

Annual reduction in pollutants are also evaluated in renewable-based energy systems and a stand-alone power system with diesel generator sets.

List of Symbols and Indices

P_{pv}	Power output of PV array (kW)
P_{pv}^r	Rated capacity of PV array (kW)
f_{pv}	PV derating factor (%)
G_T	Solar radiation on PV array (kW/m²)
$G_{T,STC}$	Solar radiation at standard test conditions (1 kW/m²)
α_p	Temperature coefficient of power (%/°C)
T_c	PV cell temperature (°C)
$T_{c,STC}$	PV cell temperature at standard test conditions (25 °C)
T_a	Ambient temperature (°C)
$T_{c,NOCT}$	Nominal operating PV cell temperature (°C)
$T_{a,NOCT}$	Ambient temperature at which NOCT defined (20 °C)
$G_{T,NOCT}$	Solar radiation at which NOCT defined (0.8 kW/m²)
η_{mp}	Efficiency of PV array at MPP (%)
σ_{BES}	Battery self-discharge rate
$E_{ch}(t)$	Battery charging energy
$E_{dch}(t)$	Battery discharging energy
η_{Conv}	Efficiency of converter
R_{batt}	Battery storage system life (years)
N_{batt}	Number of batteries
$Q_{lifetime}$	Lifetime throughput (kWh)
Q_{thrpt}	Annual storage throughput (kWh/yr)
$R_{batt,f}$	Storage system float life (years)
$C_{ref,batt}$	Replacement cost of the storage bank ($)
C_{bw}	Battery wear cost ($/kWh)
ηrt	Storage roundtrip efficiency
INV_{cap}	Rating of inverter (kVA)
L_{ind}	Total inductive load (kW)
L_0	Total non-inductive load (kW)
EC_{DG}^t	Emission cost of diesel generator
k_{DR}	Incentive rate ($/kW)
$P_{b,t}^{DR}$	Load shifted at bus 'b' and time 't'
$\propto_{DR}$	Reduction factor of load
$P_{D\,b,t}$	Load at bus 'b' and time 't'
$D_{unmet,t}$	Total unmet energy demand
D_t	Total energy demand

Role of Energy Storage Systems in the Micro-Grid Operation   239

v	Wind velocity in m/s
k	Shape parameter
c	Scale parameter (m/s)
$\bar{v}$	Mean wind speed (m/s)
σ_v	Standard deviation of the wind speed in m/s
s	Solar irradiance in W/m²
$f_b(s)$	Beta distribution function of s
α, β	Parameters of beta distribution
$C_g(t)$	Main grid electricity price
$P_g(t)$	Power import from main grid at time t
$P_i(t)$	Output power of the controllable unit i at time t
NG	Total number of controllable units
ns	Total number of scheduling time intervals
FC_i	Fuel cost of unit i
S_i	Start-up cost of unit i
EF_{ij}	Emission factor of unit i related to emission type j (SO_2, CO_2, NO_x)
EF_{gj}	Average emission factor of the main grid related to emission type j (SO_2, CO_2, NO_x)
	Total number of PV buses in the micro-grid network
P_{WT}	Wind turbine power output
P_{PV}	Photo voltaic system power output
P_{MT}	Micro turbine power output
P_{FC}	Fuel cell power output
P_{ch}	Battery charging power
P_{dc}	Battery discharging power
P_d	Load demand
SOC_i	State of charge
η_i	Charging or discharging efficiency
C_i	Capacity of the energy storage at ith unit
$\mu_{PL,i}, \mu_{QL,i}$	Mean value of active and reactive power demand
$\sigma_{PL,i}, \sigma_{QL,i}$	Standard deviation of active and reactive power demand
Pg_i, Qg_i	Real and reactive power generations
E(i)	Self-elasticity
E(i,j)	Cross elasticity
$pd_o(i)$	Initial load demand
pdr(i)	Modified load demand after implementation of DR program
ρ(i)	Spot electricity price
$ρ_o(i)$	Initial electricity price
A(i)	Incentive of DR program at ith hour
pen(i)	Penalty of DR program at ith hour

References

1. David Wenzhong Gao, *Energy Storage for Sustainable Microgrid*, 1st Edition, Elsevier, 2015.
2. J. A. Pecas Lopes, C. L. Moreira, and A. G. Madureira, "Defining control strategies for Micro Grids islanded operation," *IEEE Transaction on Power Systems*, vol. 21, no. 2, pp. 916-924, 2006.
3. Paul Breeze, *Power System Energy Storage Technologies*, Elsevier, 2018.
4. Garrett Fitzgerald, James Mandel, Jesse Morris, Herve Touati, "The Economics of Battery Energy Storage: How multi-use, customer-sited batteries deliver the most services and value to customers and the grid," Rocky Mountain Institute, September 2015. <<http://www.rmi.org/electricity_battery_value>>
5. Ibrahim Alsaidan, Amin Khodaei, and Wenzhong Gao, "A Comprehensive Battery Energy Storage Optimal Sizing Model for Microgrid Applications," *IEEE Transactions on Power Systems*, vol. 33, no. 4, pp. 3968-3980, 2018.
6. J. W. Shim, G. Verbic, K. An, J. H. Lee, and K. Hur, "Decentralized operation of multiple energy storage systems: SOC management for frequency regulation," *2016 IEEE International Conference on Power System Technology, POWERCON* 2016, pp. 1–5, 2016.
7. N. L. Díaz, A. C. Luna, J. C. Vasquez, and J. M. Guerrero, "Centralized Control Architecture for Coordination of Distributed Renewable Generation and Energy Storage in Islanded AC Microgrids," *IEEE Transactions on Power Electronics*, vol. 32, no. 7, pp. 5202–5213, 2017.
8. L. Wang, D. H. Liang, A. F. Crossland, P. C. Taylor, D. Jones, and N. S. Wade, "Coordination of Multiple Energy Storage Units in a Low-Voltage Distribution Network," *IEEE Transactions on Smart Grid*, vol. 6, no. 6, pp. 2906–2918, Nov 2015.
9. K. Chua, Y. S. Lim, P. Taylor, S. Morris, and J. Wong, "Energy storage system for mitigating voltage unbalance on low-voltage networks with photovoltaic systems," *IEEE Transactions on Power Delivery*, vol. 27, no. 4, pp. 1783–1790, 2012.
10. M. Zeraati, M. E. Hamedani Golshan, and J. Guerrero, "Distributed Control of Battery Energy Storage Systems for Voltage Regulation in Distribution Networks with High PV Penetration," *IEEE Transactions on Smart Grid*, pp. 1–1, 2016.
11. Ahmed T. Elsayed, Christopher R. Lashway, and Osama A. Mohammed, "Advanced Battery Management and Diagnostic System for Smart Grid Infrastructure," *IEEE Transactions on Smart Grid*, vol. 7, iss. 2, pp. 897-905, 2016.
12. International Energy Agency: World Energy Outlook 2019 (www.iea.org/weo).
13. N. D. Hatziargyriou, T. S. Karakatsanis, and M. Papadopoulos, "Probabilistic load flow in distribution systems containing dispersed wind power generation," *IEEE Transactions on Power Systems*, vol. 8, no. 1, pp. 159-165, 1993.

14. D. Q. Hung, N. Mithulananthan, and K. Y. Lee, "Determining PV Penetration for Distribution Systems With Time-Varying Load Models," *IEEE Transactions on Power Systems*, vol. 29, no. 6, pp. 3048-3057, 2014.
15. A. Zakariazadeh, S. Jadid, and Siano, "Smart microgrid energy and reserve scheduling with demand response using stochastic optimization," *International Journal of Electrical Power & Energy Systems*, vol. 63, 523-533, 2014.
16. Naveen Venkatesan, Jignesh Solanki, and Sarika Khushalani Solanki, "Residential Demand Response model and impact on voltage profile and losses of an electric distribution network," *Applied Energy*, vol. 96, pp. 84–91, 2012.
17. https://www.nrel.gov/gis/
18. IEEE Std 2030.2.1-2019, IEEE Guide for Design, Operation, and Maintenance of Battery Energy Storage Systems, both Stationary and Mobile, and Applications Integrated with Electric Power Systems.
19. IEEE Std 1547.4-2011, IEEE Guide for Design, Operation, and Integration of Distributed Resource Island Systems with Electric Power Systems.

9

Role of Energy Storage System in Integration of Renewable Energy Technologies in Active Distribution Network

Vijay Babu Pamshetti[1]* and Shiv Pujan Singh[2]

[1]*Department of Electrical and Electronics Engineering, B V Raju Institute of Technology, Narsapur, Medak, Telangana, India*
[2]*Electrical Engineering, Indian Institute of Technology BHU Varanasi, Uttar Pradesh, India*

Abstract

This chapter explores the role of an energy storage system (ESS) in integration of renewable energy technologies (RET) in active distribution networks (ADN). To do so, a new two-stage coordinated optimization model has been introduced for integrated planning of RET and ESS. In stage 1, optimal allocation of RET and ESS devices has been carried out simultaneously, whereas stage 2 performs optimal scheduling of active and reactive power of RET and BESS including volt/Var control (VVC) devices such as shunt capacitor banks (CBs) and on-load tap changer (OLTC) for active power optimization and voltage regulation. The objective of the proposed methodology is to minimize the overall total investment and operating cost of RET and ESS over planning horizon. Besides, it includes cost of purchased power from grid, cost of energy not served (ENS) and cost of CO_2 emission. Meanwhile, to address high-level uncertainties related to renewable energy generation and load demands, a stochastic module has been adopted, which produces a number of scenarios by using Monte Carlo simulation (MCS). Further, K-means clustering method has been adopted to determine the reduced scenarios with high quality and diversity. The proposed framework has been implemented on IEEE 33-bus distribution system for different cases and solved by using proposed hybrid optimization solver, which is performed in MATLAB-GAMS environment. The test results validate the effectiveness of the proposed framework in improving the

---

Corresponding author: vijaybabu212@gmail.com

system efficiency, enhancing the reliability and reducing carbon emission footprint of ADN, compared with the traditional planning schemes.

Keywords: Active distribution network, conservation voltage reduction, energy storage system, renewable energy technologies

Nomenclature

i, j	Index of node/bus
$\Omega_{wd}, \Omega_{pv}, \Omega_{ess}$	Set of buses installed with wind, pv and ESS respectively
Nbus	Total number of buses
N_y	Number of planning year
D	Discount rate
P_{PV}	Photovoltaic output power
P_{STD}	Power under standard test conditions (STD)
ξ	Solar irradiance (W/m2)
T_{cell}	Cell temperature (°C)
δ	Power-temperature coefficient (%/°C)
T_{amb}	Ambient temperature (°C)
OCT^{nom}	Nominal operating cell temperature conditions (°C)
$P_{WG,rated}$	Wind generation rated power
$v_{rated}, v_{ci}, v_{co}$	rated wind speed, cut-in wind speed and cut-out wind speed respectively
μ_p, μ_q	Mean value of active (p) and reactive (q) power load respectively
σ_p, σ_q	Standard deviation of active (p) and reactive (q) power load respectively
P_L, Q_L	Active (p) and reactive (q) power load respectively
P_L^{nom}, Q_L^{nom}	Nominal active (p) and reactive (q) power load respectively
k^p, k^q	Exponent component of active (p) and reactive (q) power load respectively
V, Vnom	Rated voltage and nominal voltage respectively
$SOC_{i,s,y}^t$	ESS State of charge (SOC)
SOC^{min}, SOC^{max}	Minimum and maximum limit of SOC of ESS respectively
$P_{ess,i,s,y}^t, Q_{ess,i,s,y}^t$	active (p) and reactive (q) power injected by ESS at ith bus for sth scenario in the yth year respectively
$P_{ess,i,s,y}^{ch,t}, P_{ess,i,s,y}^{dch,t}$	Active power charging and discharging by ESS at ith bus for sth scenario in the yth year respectively

Δt	Interval time
$P_{ess,i}^{ch,min}$, $P_{ess,i}^{ch,max}$	Minimum and maximum limit of active power charging of ESS at ith bus respectively
$P_{ess,i}^{dch,min}$, $P_{ess,i}^{dch,max}$	Minimum and maximum limit of active power discharging of ESS at ith bus respectively
η^{ch}, η^{dch}	charging and discharging efficiency of ESS respectively
$S_{ess,i}^{rated}$	Rated MVA power capacity of ESS at ith bus
$C_{inv,y}^{tot}$	Total investment cost in the yth year
$C_{pv}^{inv}, C_{wd}^{inv}$	investment cost of PV and wind generation respectively
$P_{wd,i}^{rated}, P_{pv,i}^{rated}$	Rated power capacity of wind and PV generation at ith bus respectively
$C_{P,ess}^{inv}, C_{E,ess}^{inv}$	investment cost of ESS power and energy capacity respectively
$P_{ess,i}^{rated}, E_{ess,i}^{rated}$	Rated power and energy capacity of ESS at ith bus respectively
A^{wd}, A^{pv}, A^{bes}	Annual equivalent factor of wind, PV generation and ESS respectively
$C_{om,y}^{tot}, C_{ens,y}^{tot}$	Overall operation and maintenance (O&M) cost and cost of energy not served respectively
$C_{sub,y}^{tot}, C_{emis,y}^{tot}$	Overall cost of power taken from substation and cost of CO_2 emission respectively
N_t	Total number of time periods
$N_{s,y}$	Total number of scenarios in the yth year
$N_{s,r}$	Total number of reduced scenarios in the yth year
$N_{d,y}$	Total number of days in the yth year
$Prob_{k,s,y}$	Probability of kth variable for sth scenario in the yth year
$\pi_{s,y}$	Probability of normalised sth scenario in the yth year
$C_{wd}^{OM}, C_{pv}^{OM}, C_{ess}^{OM}$	O&M cost of wind, PV generation and ESS respectively
$P_{wd,i,s,y}^{t}, Q_{wd,i,s,y}^{t}$	Active and reactive power injected by wind generation at ith bus for sth scenario in the yth year respectively
$P_{pv,i,s,y}^{t}, Q_{pv,i,s,y}^{t}$	Active and reactive power injected by PV generation at ith bus for sth scenario in the yth year respectively
$C_{s,y}^{sub,t}, C_{s,y}^{emis,t}, C_{s,y}^{ens,t}$	Cost of power taken from substation and cost of CO_2 emission at tth hour for sth scenario in the yth year respectively
$P_{sub,s,y}^{t}$	Active power taken from substation for sth scenario in the yth year respectively
$P_{ens,s,y}^{t}$	Energy not served for sth scenario in the yth year respectively

$P_{wd,i,y}^{max}, P_{pv,i,y}^{max}, P_{ess,i,y}^{max}$ Active power injection of wind, PV and ESS at ith bus for sth scenario in the yth year respectively

$P_{i,s,y}^{t}, Q_{i,s,y}^{t}$ Active and reactive power injected at ith bus for sth scenario in the yth year respectively

$V_{i,s,y}^{t}$ voltage magnitude at ith bus for sth scenario in the yth year

$G_{ij,y}, B_{ij,y}$ Conductance and susceptance of branch in the yth year respectively

V_i^{min}, V_i^{max} Minimum and maximum limit of voltage magnitude respectively

$S_{l,s,y}^{t}$ MVA at lth branch for sth scenario in the yth year

S_l^{rated} rated MVA capacity of lth branch

$\varphi_{wd,i,s,y}^{t}$ power factor of wind generator at ith bus for sth scenario in the yth year

$\varphi_{wd,i}^{min}, \varphi_{wd,i}^{max}$ Minimum and maximum limit of power factor of wind generator at ith bus respectively

$Q_{pv,i}^{max}$ Maximum limit of reactive power of photovoltaic inverter

tap_{oltc}^{t} On-load tap changer (OLTC) transformer tap position at tth hour

$tap_{oltc}^{min}, tap_{oltc}^{max}$ Minimum and maximum limit of OLTC transformer tap position respectively

9.1 Introduction

9.1.1 Background

Due to three key factors—environmental concerns, technological innovation and government policies—the integration of renewable energy sources (RES) has been increasing tremendously all over the world. In the period 2010-2019, about $2.6 trillion was invested in renewable energy capacity, triple the amount financed in the previous decade [1]. Figure 9.1 depicts that the solar generation, wind generation, and biomass and waste-to-energy have funded the $1.3 trillion, $1 trillion, and $115 billion, respectively.

Figure 9.2 depicts the amount invested in renewable energy sources (RES) in 20 markets. More than $14 billion has spent on RES in each nation. China has invested about $758 billion, approximately 31% of the global total, whereas the U.S. has invested about $356 billion.

In renewable energy development in India [2], as per July 2017, the total RES installed was 58.9 GW, superior that of hydro-power generation,

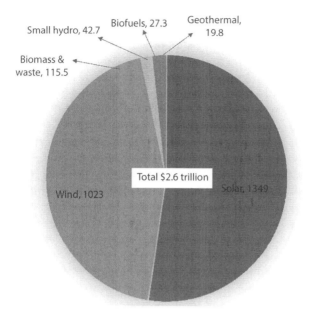

Figure 9.1 Pie chart of RES Investment in the period 2010-2019, $Billions.

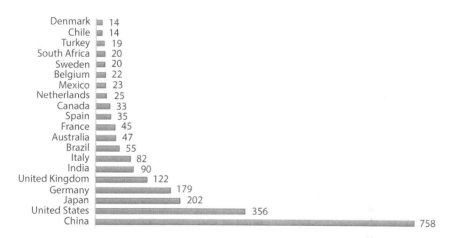

Figure 9.2 Bar diagram of RES Investment in the period 2010 to 1st half of 2019, $Billions.

which was 44.6 GW. The share of RES in the total installed capacity also increased to 14%, 16%, and 17.8% in March 2016, April 2017, and July 2017, respectively. On the other side, the availability of advanced RES technologies and their diminishing trend of costs, along with the complicated nature of building new transmission lines, the increasing demand for higher reliability of supply, among others factors, has encouraged significant investment in RESs.

9.1.2 Motivation and Aim

As demonstrated in the background, RESs will play a key role in distribution systems as per environmental concerns. However, the uncertainty and intermittent nature of RES often poses several techno-economic challenges in the system. Besides, high penetration of RESs will inescapably involve adjusting and shifting the existing infrastructure.

The arbitrary allocation and high integration of RES units in the system exacerbate problems such as voltage rise, reversible power flow, and uncertainty associated with power output. Therefore, RES allocation should be prudently planned to maximize system efficiency. Besides, the intermittent and uncertain nature of RES (i.e., wind and solar generation) and load can cause sudden voltage variation in distribution networks. Further, the conventional volt/VAR control strategies in distribution systems are expected to face numerous challenges, which leads to over-voltage, under-voltage, increasing system losses, and unnecessary wear and tear of VVC devices. Therefore, it is necessary to mitigate volt/VAR control issues and facilitate a seamless integration for large penetration of RES, which are needed to coordinate with traditional utility VVC devices.

An energy storage system (ESS) is a key component to ensure reliable integration of renewable-based generation, offer regulation capabilities at the distribution-transmission interface, and provide reliable power supply in both urban and rural environments. ESS may provide unique advantages over conventional resources to utilities and their customers, including rapid and accurate response and ramping as flexible capacity, reliability, modularity, and locational flexibility. If storage can be optimized for the right location, size, and duration along with being produced at a very low cost, then load can be scheduled to match the generation profile so as to reduce the overall cost of energy. This will thereby introduce a significant paradigm shift in the conventional operational philosophy of generation being scheduled to match load. ESS technologies have great potential to revolutionize grid operation and control. However, it is necessary to perform feasibility and screening studies that not only optimize the size and location of storage but also identify the best grid services that justify the high capital investment costs for commercially proven ESS technologies. Hence, proper planning and operation of ESS in integration of RET in active distribution network is crucial for enhancing the performance of distribution system.

9.1.3 Related Work

A number of methods have been proposed for the optimal placement of RES in the distribution networks to minimize losses and also improve the voltage profile of the system. Based on exact power loss formula, an analytical method was proposed by Acharya et al. [3] to determine the optimal allocation of distributed generator (DG). Hung et al. [4, 5] introduced an analytical method to determine the location and sizes of multiple DGs in a distribution system. In [6] an analytical method was used by Wang and Nehrir to decide the location of optimally placed DG in radial as well as meshed distribution networks in order to minimize real power loss. A multi-objective evolutionary algorithm was employed by Celli et al. [7] for sizing and sitting of DG in the distribution systems.

Lee and Park [8] have employed an Kalman filter algorithm for determining the optimal site and sizes of distributed energy resources (DERs) in order to minimize the power loss. Their work [9] has recently been extended to introduce a optimal locator index (OLI) factor to identify the optimal locations. A multi-objective mixed integer formulation and its solution by genetic algorithm (GA) has been employed by Shaaban et al. [10] considering the uncertainty and variability nature of RES and loads. Naik et al. [11] suggested an analytical method to determine the optimal location and capacities of DERs to minimise the power losses. Ameli et al. [12] applied a Multi-objective particle swarm optimization to determine the optimal locations and sizes of DG considering operational and economic aspects.

In [13] power flow solution method was proposed by Elsaiah et al. and further extended to determine the optimal allocation of DERs in distribution system for loss reduction. Comprehensive analysis of different loss sensitivity methods for DER placement has been presented by Murthy and Kumar [14]. A thumb rule called 2/3 rule was developed by Willis [15] for optimal DER placement in the distribution system. In [16] based on algebraic equations, an analytical approach was proposed by Muttaq et al. to determine the optimal operation, size and location of the DER. Deepak et al. [17] employed a cat swarm optimization, for optimal allocation of DERs in the distribution networks.

Esmaili [18] used application of GAMS to solve a nonlinear programming (NLP) for optimal DG placement. In [19], based on exact loss formula GA Shukla et al. determined the optimal location of DG. Combination of Particle swarm optimisation (PSO), and gravitational search algorithm (GSA) was applied by Tan et al. [20] to solve the optimal location and size of DERs in the distribution system. Georgilakis et al. [21] presented a rigorous review of the state-of-the-art models and optimization techniques employed to the DER placement problem. Naveen et al. [22] have employed PSO considering constriction factor for the DER planning in the distribution networks.

In the last two decades, several DGs planning schemes in active distribution network (ADN) have been proposed. In [23–27] the authors depicted the significance of simultaneously planning of RES and ESS. In [23], cooperative planning model of DERs was proposed, with the objectives of reduction of investment and operating costs of the distribution network. In [24] a multi-stage planning model was proposed for simultaneous allocation of RES and ESS. In [25], energy storage has been deployed for maximum utilization of RES in ADN. Benefits of coordinated allocation and scheduling of BES system in the high PV penetrated distribution system has been presented in [26]. In [27], optimal DER allocation problem was solved by GA considering the uncertainties of the RES and load. However ref. [23–27], do not consider the benefits of Volt-VAR control devices inplanning of DG.

In order to improve the controllability and flexibility to the operation of ADN, Ref. [28–37] incorporate the Volt-VAR control techniques in planning problem. Ref [28] presented the benefits of conservation voltage reduction (CVR) operation in the planning of ESS in high penetrated RES. Optimal planning model of ESS in ADN considering soft open point (SOP), distribution network reconfiguration (DNR) and reactive power capability of DG has been proposed in [29]. Similarly, in [30] coordinated methodology for allocation of DGs, capacitor banks and SOPs has been proposed. Ref [31] demonstrated the role of oversizing of utility-owned renewable DG smart inverter considering VVC devices. Ref. [32] proposed the various planning schemes based on photovoltaic smart inverter (PVSI) control for significant benefits in enabling the large-scale integration of RES. Ref [33] presented the benefits of DNR operation in the planning of EES with high penetrated RES. Ref [34] reveals the significance of DNR and demand-side management while DG planning in active distribution network. Ref [35] demonstrated the significance of CVR operation in the formulation of planning problem. Similarly, ref [36] studied implementation of CVR operation in planning of DGs and ESS simultaneously. Ref [37] reported the importance of the VVC devices such as OLTC, and CB, while DERs planning in ADN. Studies in references [28–37] reveal the significance of DNR and CVR techniques in planning of DER. However, combined impact of DNR and CVR techniques considering advanced power electronics-based devices such as smart inverter and SOPs have not been considered in these studies. Further, Table 9.1 presents the summary of recent DER planning techniques. Thus, a highly efficient and effective

Two-Stage Coordinated Optimization Model   251

Table 9.1 Literature survey.

Reference	Published year	Decision variables	Uncertainty handling	Voltage dependent load model	DG/RES	ESS	Advanced devices	Loss	Reliability	Emission	VVC	Formulation	Solution method
[23]	2019	Size, location, type	Y	N	Y	Y	N	Y	Y	Y	N	MINLP	PSO
[25]	2018	Size, location, type	Y	N	Y	Y	N	Y	N	N	N	MINLP	ant lion optimizer (MOALO)
[26]	2018	Size, location	N	N	Y	Y	N	Y	N	N	N	MINLP	GA
[27]	2013	Size, location	Y	N	Y	N	N	Y	N	N	N	MINLP	GA
[29]	2018	Size, location	Y	N	Y	Y	Y	Y	N	N	N	MISOCP	MOSEK solver
[30]	2018	Size, location, type	Y	N	Y	N	Y	Y	N	N	N	MINLP	GA
[31]	2019	Size	Y	N	Y	N	Y	Y	N	N	N	NLP	KINTRO
[34]	2018	Size, location, type	Y	N	Y	N	Y	Y	N	N	N	MINLP	Differential evolutionary algorithm
[37]	2019	Size, location, type	Y	N	Y	Y	N	Y	Y	N	N	MISOCP	CPLEX
[38]	2009	Size, location	N	Y	Y	N	N	Y	N	N	N	MINLP	GA

(*Continued*)

Table 9.1 Literature survey. (Continued)

Reference	Published year	Decision variables	Uncertainty handling	Voltage dependent load model	DG/RES	ESS	Advanced devices	Loss	Reliability	Emission	VVC	Formulation	Solution method
[39]	2012	Size, location, type	Y	N	Y	N	N	Y	Y	N	N	MINLP	GA
[40]	2012	Size, location	N	N	Y	N	N	Y	N	N	N	MINLP	GA/PSO
[41]	2015	Size, location, type	Y	N	Y	N	N	Y	N	N	N	MILP	CPLEX
[42]	2018	Size, location, type	Y	N	Y	N	N	Y	Y	Y	N	MILP	CPLEX
[43]	2019	Size, location	Y	N	Y	N	Y	Y	N	N	N	MILP	CPLEX
[44]	2018	---	N	N	Y	Y	Y	Y	N	N	N	MINLP	GA
[45]	2014	Size, location	N	Y	Y	N	N	Y	N	N	N	MINLP	Analytical method
[46]	2017	Size, location, type	Y	N	Y	Y	N	Y	Y	Y	N	MILP	MILP solver

Y-Yes considered. N-Not considered.

approach for optimal coordination of the DER planning and their scheduling in conjunction with advanced techniques is highly desired to enhance performance of distribution systems.

9.1.4 Main Contributions

It can be comprehended from the above review that the role of ESS in integration of RET considering advanced CVR scheme has not been fully explored so far. In view of this, the present paper proposes a new integrated planning of RET and ESS devices in ADN considering advanced CVR scheme. The contribution of the present work is as follows.

- ➢ Proposed a new two-stage coordinated optimization framework for integrated operational and planning model of RET and ESS in ADN.
- ➢ Proposed an integrated long-term planning model of RET and ESS that addresses the economical, operational, and environmental issues. The uncertainties of renewable generation load demand are entirely taken into account by using a stochastic module.
- ➢ Impact of CVR technique on RET and ESS planning in ADN has been studied.
- ➢ Proposed a seamless method for secure coordination and integration of DERs at a large scale in order to enhance reliability, flexibility, power quality, and resiliency.
- ➢ Introduced a hybrid optimization solver to solve the large-scale non-convex nonlinear mixed integer programming problem without linearization or relaxation.

The proposed algorithm has been validated on IEEE 33 bus distribution system.

9.2 Active Distribution Network

Due to a number of technical and ecofriendly aspects, there has been a growing interest in integrating distributed energy resources (DER) in the distribution networks. Hence, the conventional passive distribution networks will evolve into active distribution network (ADN) by employing DERs such as wind generation, photovoltaic generation, energy storage systems (ESS), and electrical vehicle station (EVS), as shown in Figure 9.3. On the other

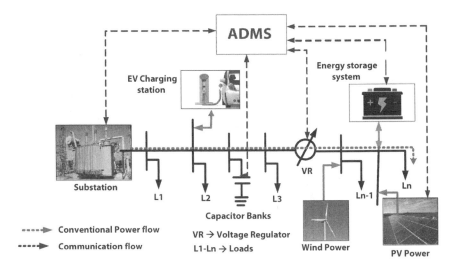

Figure 9.3 Schematic of active distribution system with ADMS.

hand, technological advances have enabled distributed management systems (DMSs), which offer a suite of applications such as Volt/VAR Optimization (VVO), Fault Location, Isolation, and Service Restoration (FLISR), conservation voltage reduction (CVR), Optimal Power Flow (OPF), and Outage Management System (OMS) etc. which are planned to monitor and control the distribution networks proficiently and consistently.

Key to enabling a transformative change in future distribution grids is to enable a seamless and secure coordination and integration of DERs at a large scale to enhance reliability, power quality, and resiliency, while minimizing capital investment and operating costs. Coordination of pervasively installed DERs provides opportunities for reliable and sustainable electricity delivery in both urban and rural areas.

9.3 Uncertainties Modelling of Renewable Energy Sources and Load

9.3.1 Uncertainty of Photovoltaic (PV) Power Generation

In this chapter, Beta distribution function has been adopted to exhibit the uncertainty of solar irradiance [34] and probability density function (PDF) is given in (9.1)

$$PDF_s(\xi) = \begin{cases} \dfrac{\Gamma(\alpha+\beta)}{\Gamma(\alpha)\Gamma(\beta)} \xi^{(\alpha-1)}(1-\xi)^{(\beta-1)}; & 0 \le \xi \le 1, \alpha, \beta \ge 0 \\ 0 & \text{otherwise} \end{cases} \quad (9.1)$$

where, $\Gamma(\bullet)$ is the Gamma function, ξ is the solar irradiance, α and β are function parameters and determined from the mean (μ) and standard deviation (σ) of solar irradiance ξ as given below

$$\beta = (1-\mu)\left(\dfrac{\mu(1+\mu)}{\sigma^2} - 1\right); \quad \alpha = \dfrac{\mu \times \beta}{1-\mu} \quad (9.2)$$

Power outputs of each PV module is a function of solar irradiance, ambient temperature and physical characteristics of each module. The output power in PV module can be determined by the power characteristic model as given in (9.3).

$$P_{pv} = P_{STD}\left\{\dfrac{\xi}{1000}[1+\delta(T_{cell} - 25)]\right\} \quad (9.3)$$

$$T_{cell} = T_{amb} + \left(\dfrac{OCT^{nom} - 20}{800}\right)\xi \quad (9.4)$$

9.3.2 Uncertainty of Wind Power Generation

Weibull distribution has been adopted to exhibit the uncertainty of wind speed [34] and probability density function (PDF) is shown in (9.5)

$$PDF_w(V) = \left(\dfrac{k}{c}\right)\left(\dfrac{v}{c}\right)^{(k-1)} \exp\left(-\left(\dfrac{v}{c}\right)^k\right) \quad (9.5)$$

Where k, c are Weibull parameters. v is the wind speed.

Power outputs of wind generation (WG) connected to wind turbine is a function of wind speed and physical characteristics of turbine. The output power in the wind turbine (WT) can be calculated by using (9.6) and (9.7) respectively.

$$P_{WG} = \begin{cases} 0; & 0 \le v < v_{ci} \\ a + b \times v^3; & v_{ci} \le v < v_{rated} \\ P_{WG,rated}^n; & v_{rated} \le v < v_{co} \\ 0; & v_{co} \le v \end{cases} \quad (9.6)$$

$$a = \left(P_{WG,rated}^n \times v_{ci}^3 \right) / \left(v_{ci}^3 - v_{rated}^3 \right)$$
$$b = P_{WG,rated}^n / \left(v_{rated}^3 - v_{ci}^3 \right)$$
(9.7)

9.3.3 Voltage Dependent Load Modelling (VDLM)

Most of the loads in the distribution network exhibit the voltage dependent behavior [47]. These loads are highly reliant on the voltage magnitude. There are two kinds of VDLMs, namely exponential load model (ELM) and polynomial load model (PLM). In this chapter, ELM have been selected to represent the load-to-voltage sensitivities. The active and reactive power load are determined by using (9.8) and (9.9) respectively.

$$P_L = P_L^{nom} \left(\frac{V}{V^{nom}} \right)^{k^p} \quad (9.8)$$

$$Q_L = Q_L^{nom} \left(\frac{V}{V^{nom}} \right)^{k^q} \quad (9.9)$$

Normal distribution function has been employed to exhibit the uncertainty of loads [34] and active and reactive power load are given in equation (9.10) and (9.11) respectively.

$$PDF_{pl}(P_L) = \frac{1}{\sqrt{2\pi}\sigma_p} \exp\left[-\frac{(P_L^{nom} - \mu_p)^2}{2\sigma_p^2} \right] \quad (9.10)$$

$$PDF_{ql}(Q_L) = \frac{1}{\sqrt{2\pi}\sigma_q} \exp\left[-\frac{(Q_L^{nom} - \mu_q)^2}{2\sigma_q^2} \right] \quad (9.11)$$

9.3.4 Proposed Stochastic Variable Module for Uncertainties Modelling

Stochastic Variable Module (SVM) has been introduced in order to model the uncertainties of variables such as wind speed, solar irradiance and electrical load as illustrated in Figure 9.4.

Historical data (i.e., wind speed, solar irradiance and load) have been collected for a specified time period and determine the mean and variance of each variable with the help of the collected data. A large number

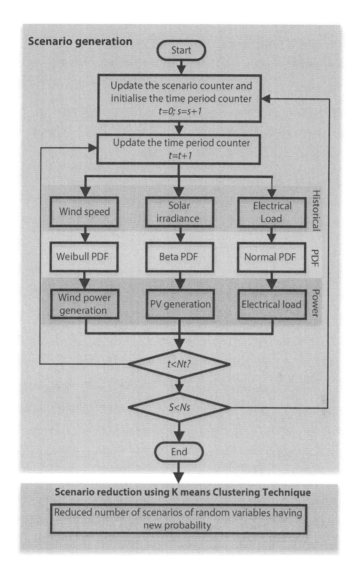

Figure 9.4 Stochastic variable module (SVM) for uncertainties.

of scenarios have been generated with the help of probability distribution function (PDF) using the Monte Carlo simulation (MCS) method within specified mean and variance. As described in above Section 9.3.3, Weibull distribution, Beta distribution and Normal distribution have been employed to exhibit the uncertainties of wind speed, solar irradiance and load, respectively [34]. Then the wind speed and solar irradiance are converted into wind power generation and PV power generation based on the wind turbine performance curve and PV module characteristics, respectively. It is important to mention that as the number of scenarios increases, the computational time will also increase. The distribution system operator (DSO) must decide the settings of control variables as as fast as possible, to accommodate the hourly operation schedule of a day. Therefore, an effective scenario reduction technique is required to reduce the scenarios with high quality and diversity. For this purpose, the K-mean clustering technique [48] has been adopted to reduce the number of scenarios. The purpose of this technique is to arrange original scenarios of solar irradiance, wind speed and loads into clusters according to their likenesses. The centroid of each cluster is identified as the mean value of the wind speed, solar irradiance and loads. After implementing the K-mean clustering technique, the probability of the reduced scenarios is normalized using expression (9.12).

$$\pi_s = \frac{\Pi_{k=1}^{N_v} \operatorname{Prob}_{k,s}}{\sum_{s=1}^{N_{s,r}} \Pi_{k=1}^{N_v} \operatorname{Prob}_{k,s}} \tag{9.12}$$

9.3.5 Modelling of Energy Storage System

The following constraints (9.13)-(9.19) have been incorporated in the optimization to represent the model of ESS in the steady-state analysis.

$$SOC_{i,s,y}^t = SOC_{i,s,y}^{t-1} + P_{ess,i,s,y}^t \times \Delta t \tag{9.13}$$

$$SOC^{\min} \leq SOC_{i,s,y}^t \leq SOC^{\max} \Delta \tag{9.14}$$

$$SOC_{i,s,y}^{t=24} = SOC_{i,s,y}^{t=1} \tag{9.15}$$

$$\left. \begin{array}{c} P_{ess,i}^{ch,\min} \leq P_{ess,i,s,y}^{ch,t} \leq P_{ess,i}^{ch,\max} \\ P_{ess,i}^{dch,\min} \leq P_{ess,i,s,y}^{dch,t} \leq P_{ess,i}^{dch,\max} \end{array} \right\} \tag{9.16}$$

$$P_{ess,i,s,y}^{ch,t} \times P_{ess,i,s,y}^{dch,t} = 0 \tag{9.17}$$

$$P_{ess,i,s,y}^t = \eta^{ch} \times P_{ess,i,s,y}^{ch,t} - P_{ess,i,s,y}^{dch,t}/\eta^{dch} \tag{9.18}$$

$$Q_{ess,i,s,y}^t \leq \sqrt{\left(S_{ess,i}^{rated}\right)^2 - \left(P_{ess,i,s,y}^t\right)^2} \tag{9.19}$$

Constraints (9.13) and (9.14) represent the status of state of charge (SOC) at a particular time and node, and SOC limits of ESS, respectively. Equation (9.15) regulates the energy remaining in the ESSs at the last time interval equal to the initial energy. Constraint (9.16) represents the limits of active power charging and discharging of ESS. Equation (9.17) ensures that only one state (i.e., charging, discharging or no action) remain active at each time interval. Equation (9.18) defines the active power of ESS power based on charging and discharging power with consideration of efficiency. Constraint (9.19) represents the reactive power injection/absorption constraint of ESSs converter.

9.3.6 Basic Concept of Conservation Voltage Reduction

CVR is a serviceable method adopted by distribution system operators for energy savings. The principle of CVR is "to achieve energy savings by gradually decreasing the voltage magnitude at the distribution transformer secondary terminal end, which ensures that the voltage magnitude of all nodes in the distribution feeders are within permissible voltage limits. According to the American National Standards Institute (ANSI), the permissible range of voltage magnitude can be set as ±5% of the nominal value" [49].

9.3.7 Framework of Proposed Two-Stage Coordinated Optimization Model

The proposed integrated planning model not only determines the location, capacity and type of RET and ESS but also simulates the optimal active and reactive management of RET and ESS. The formulation of the proposed model in two stages corresponding to the integrated planning and operation model, respectively, has been shown in Figure 9.5.

In stage 1 (i.e., integrated planning level), the system data is fed to the integrated planning optimization model, which involves the planning of RET and ESS. In this stage, decision variables are the locations and sizes of RET and ESS to be installed. For coordination purpose, all the candidate planning proposals would be transported to the integrated operation optimization model in stage 2 (i.e., integrated operation level), where optimal scheduling of RES, BESS and OLTC devices would be performed considering uncertainties of renewable energy generation and loads. After this step, the posted scheduling of RET, ESS and OLTC devices corresponding to RET and ESS capacities would be stored and fed back to planning model in stage 1, in order to revise previous planning scheme. This process is

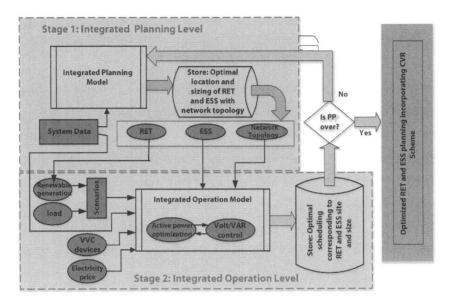

Figure 9.5 Framework of proposed two-stage coordinated optimization.

continued up to planning period (PP) considering the load growth in each period. Finally, the output file presents the optimized operational planning of RET and ESS.

9.3.8 Proposed Problem Formulation

The objective function (OF) of present optimization problem connects the minimization of the total investment and operating costs of RET and ESS devices over a planning horizons, it can be formulated as given in (9.20),

$$Minimize : OF = IC + OC \qquad (9.20)$$

where IC represents the net present value (NPV) of RET and ESS investment costs, which are determined from stage 1 (i.e., integrated planning level); OC denotes the NPV of total expected operating costs under all the possible operating scenarios. The stage 1 represents the planning level optimization for minimising the total investment cost (IC), which includes installation cost of RET and ESS devices and is given as follows:

$$IC = \sum_{y=1}^{Ny} \left[C_{inv,y}^{tot} \times (1+d)^{-y} \right] \qquad (9.21)$$

Where

$$C_{inv,y}^{tot} = \left[A^{wd} \sum_{i \in \Omega_{wd}} C_{wd}^{inv} P_{wd,i}^{rated} + A^{pv} \sum_{i \in \Omega_{pv}} C_{pv}^{inv} P_{pv,i}^{rated} \right.$$
$$\left. + A^{bes} \sum_{i \in \Omega_{ess}} \left(C_{P,ess}^{inv} P_{ess,i}^{rated} + C_{E,ess}^{inv} E_{ess,i}^{rated} \right) \right] \quad (9.22)$$

here

$$A^x = \frac{d(1+d)^{LT_x}}{(1+d)^{LT_x} - 1}; x \in \{wd, pv, ess\}$$

A^x represents the equivalent annual cost, which is used to convert the total cost into annual cost. Here, d and LT represent the discount rate and the asset life time period, respectively.

Stage 2 represents operation level optimization through repeated simulations for minimizing the expected total operating costs (OC), which includes the operational and management cost of DER and SOP, power purchased from the substation as well as the emission cost and cost of energy not served as expressed in (9.23)

$$OC = \sum_{y=1}^{N_y} \left[\frac{C_{om,y}^{tot} + C_{sub,y}^{tot} + C_{ens,y}^{tot} + C_{emis,y}^{tot}}{(1+d)^y} \right] \quad (9.23)$$

- Yearly operational and management cost of RET and ESS

$$C_{om,y}^{tot} = \sum_{s=1}^{N_{s,y}} N_{d,y} \times \pi_{s,y} \times \sum_{t=1}^{T} \left[\sum_{i \in \Omega_{wd}} C_{wd}^{OM} P_{wd,i,s,y}^{t} \right.$$
$$\left. + \sum_{i \in \Omega_{pv}} C_{pv}^{OM} P_{pv,i,s,y}^{t} + \sum_{i \in \Omega_{ess}} C_{ess}^{OM} P_{ess,i,s,y}^{t} \right] \quad (9.24)$$

- Yearly cost of energy purchased from the substation

$$C_{sub,y}^{tot} = \sum_{s=1}^{N_{s,y}} N_{d,y} \times \pi_{s,y} \times \sum_{t=1}^{T} C_{s,y}^{sub,t} \times P_{sub,s,y}^{t} \quad (9.25)$$

- Yearly cost of CO_2 emission from the substation

$$C_{emis,y}^{tot} = \sum_{s=1}^{N_{s,y}} N_{d,y} \times \pi_{s,y} \times \sum_{t=1}^{T} C_{s,y}^{emis,t} \times P_{sub,s,y}^{t} \quad (9.26)$$

- Yearly cost of energy not served

$$C_{ens,y}^{tot} = \sum_{s=1}^{N_{s,y}} N_{d,y} \times \pi_{s,y} \times \sum_{t=1}^{T} C_{s,y}^{ens,t} \times P_{ens,s,y}^{t} \quad (9.27)$$

9.3.8.1 Investments Constraints

- Budget constraint

$$IC \leq \cos t^{budget} \quad (9.28)$$

- Limits of RET penetration at each bus

$$\left. \begin{array}{l} 0 \leq P_{wd,i,s,y}^{t} + P_{pv,i,s,y}^{t} \leq P_{L,i,s,y}^{t} \\ 0 \leq P_{wd,i,s,y}^{t} \leq P_{wd,i,y}^{max} \\ 0 \leq P_{pv,i,s,y}^{t} \leq P_{pv,i,y}^{max} \end{array} \right\} \quad (9.29)$$

9.3.8.2 Operational Constraints

- Active and reactive power flow equations for each scenario

$$\left. \begin{array}{l} P_{i,s,y}^{t} = V_{i,s,y}^{t} \sum_{i=1}^{Nbus} V_{j,s,y}^{t} \left(G_{ij,y} \cos\theta_{ij,s,y}^{t} + B_{ij,y} \sin\theta_{ij,s,y}^{t} \right) \\ Q_{i,s,y}^{t} = V_{i,s,y}^{t} \sum_{i=1}^{Nbus} V_{j,s,y}^{t} \left(G_{ij,y} \sin\theta_{ij,s,y}^{t} - B_{ij,y} \cos\theta_{ij,s,y}^{t} \right) \end{array} \right\} \quad (9.30)$$

- Limits of bus voltage magnitude for each scenario

$$V_{i}^{min} \leq V_{i,s,y}^{t} \leq V_{i}^{max} \quad (9.31)$$

- Limits of branch capacity for each scenario

$$0 \leq S_{l,s,y}^{t} \leq S_{l}^{rated} \quad (9.32)$$

- Reactive power constraint of wind turbine generation (WTG) for each scenario

$$\left.\begin{array}{l} Q_{wd,i,s,y}^{t} = P_{wd,i,s,y}^{t} \tan(\varphi_{wd,i,s,y}^{t}) \\ \varphi_{wd,i}^{min} \leq \varphi_{wd,i,s,y}^{t} \leq \varphi_{wd,i}^{max} \end{array}\right\} \quad (9.33)$$

- Reactive power constraint of PV for each scenario

$$\left.\begin{array}{l} Q_{pv,i,s,y}^{t} = \sqrt{\left(S_{pv,i}^{max}\right)^2 - \left(P_{pv,i,s,y}^{t}\right)^2} \\ -Q_{pv,i}^{max} \leq Q_{pv,i,s,y}^{t} \leq Q_{pv,i}^{max} \end{array}\right\} \quad (9.34)$$

- Operation constraint of transformer

$$tap_{oltc}^{min} \leq tap_{oltc,y}^{t} \leq tap_{oltc}^{max} \quad (9.35)$$

9.3.9 Proposed Solution Methodology

The problem formulated in this chapter is a large-scale nonlinear mixed integer programming (MINLP) problem, which is difficult to solve by conventional optimization techniques. In order to overcome this problem, the problem has been viewed as two models, namely, the planning and operation models. These models have been solved using proposed two-stage framework as explained earlier in Section 9.3.7. The stage 1 and stage 2 parameters are evaluated by the decision variable sets, D1 and D2, respectively. The set D1 consists of candidate solutions for installations of RET and ESS corresponding to the network topology as illustrated in Figure 9.6(a). Similarly, set D2 consists of operational scheduling of RET, ESS, OLTC and CBs devices as illustrated in Figure 9.6(b). In this chapter, the stage 1 and stage 2 have been solved concurrently with the help of the evolutionary algorithms and GAMS optimization tool, respectively.

Figure 9.7 illustrates the implementation of proposed two-stage coordinated optimization model using MATLAB-GAMS platform. In this paper, a widely accepted particle swarm optimization (PSO) evolutionary algorithm has been employed to generate the candidate solutions D1 under MATLAB environment in order to minimize the total investment subject to the constraints given in (9.28)-(9.29). Each candidate solution (D1) obtained by stage 1 is transferred to GAMS environment through GAMS

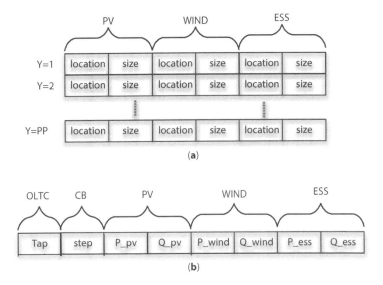

Figure 9.6 (a) Decision variable in planning stage (D1) and (b) Decision variable in operation stage (D2).

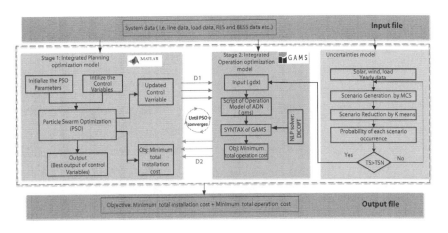

Figure 9.7 Two-stage coordinated optimization model solved by hybrid optimization solver.

data exchange (GDX) files. For each candidate solution D1, at stage 2 the problem relating the operation scheduling of OLTC taps, RET and BESS has been solved dynamically for each hour for each possible scenario. The formulated MINLP problem in stage 2 has been solved by using DICOPT, MINLP solver under GAMS environment. The MINLP optimization problem subject to the constraints given in (9.30)-(9.35). Here, the objective is to recursively minimize the daily operation costs at the candidate installation variables provided from the PSO solver. The solution of the stage 2 yields the set of decision variables for the operation scheduling, i.e., D2 for each scenario s at hour t. This process is repeated until the PSO solver converges. Further, simulation process is repeated for given planning period (pp).

9.3.10 Simulation Results and Discussions

9.3.10.1 Simulation Platform

The proposed two-stage coordinated optimization model is implemented on MATLAB-GAMS environment [50]. The optimization of formulated problem in the upper level has been solved by particle swarm optimization. Further, optimization of formulated problem in the lower level has been solved by DICOPT solver [51], non-linear programming solver of GAMS.

9.3.10.2 Data and Assumptions

The performance of the proposed methodology has been tested on a standard IEEE 33-bus distribution system [52], which is shown in Figure 9.8. The detailed data of modified IEEE 33-bus distribution system has been depicted in Table 9.2. The tap position of OLTC transformer can vary 16 steps (i.e., -8 to 8). The change in each step is 0.625%. Figure 9.9 portrays the typical load profile, power price purchased from substation, PV generation and wind generation output over a typical day. The cost parameters of PV and wind have been taken from [34]. The cost parameters of ESS have been taken from [23]. The load growth is taken as 7% for each year. The cost of unserved energy is taken as 2000 $/MWh [46]. The emission

Table 9.2 Modified 33-bus distribution system information.

Parameters	Values
Primary voltage (kV)	33
Secondary voltage (kV)	12.66
Nominal active power demand (MW)	3.715
Nominal reactive power demand (MVAR)	2.3
CBs installed locations	3, 6, 11, 23, 30, 15
Permissible voltage limits	0.95-1.05 pu

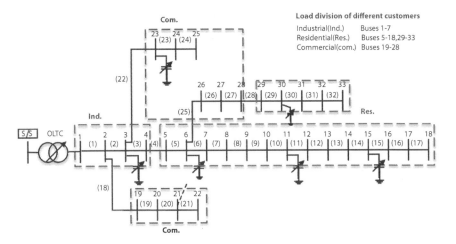

Figure 9.8 Modified 33 bus distribution system.

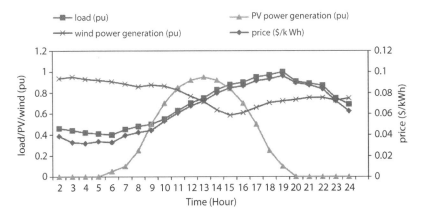

Figure 9.9 Typical load, grid price, PV and wind generation output.

rate of purchased power from grid is taken as 0.4 tCO_2e/MWh [53] and the emission price is set as 60 \$/tCO_2e [46].

9.3.10.3 Numerical Results and Discussions

The proposed optimization model has been solved considering four different cases as given in Table 9.3. Case A is corresponding to base case scenario, where no investments on RET and ESS are made. Case B corresponds to scenario, where investments on RET (i.e., PV and wind generation) are made. Case B can be referred to as traditional planning scheme. Case C is

Table 9.3 Different cases studied.

Cases	RET	ESS	CVR
Case A	No	No	No
Case B	Yes	No	No
Case C	Yes	Yes	No
Case D	Yes	Yes	Yes

similar to Case B but considers the ESS installation. Case D is similar to Case C but considers the conservation voltage reduction operation.

Table 9.4 shows the most relevant variables such as investment cost, operation & maintenance (O&M) cost of RET and ESSs, energy purchased cost from the grid, energy not served (ENS) cost and CO_2 emission cost of different cases over the five-year planning horizon.

The results of case B to Case D show a significant benefit compared to the base case (i.e., case A). The optimal planning and operation of RETs in distribution system can yield total cost savings up to \$64,37,510 (i.e., a 30.31% reduction) as seen in case B compared with that of case A. With the installation of ESS units, an additional 9% reduction in total cost savings has been achieved, although the total investment and O&M cost of RET and ESS as seen in case C under Table 9.4. Further, 1.33% more reduction has been achieved with the association of CVR operation as seen in case D.

Table 9.4 Results of different cases over planning horizon.

Cost parameters	Case A	Case B	Case C	Case D
Investment cost ($\times 10^3$ \$)	0.00	1431.89	3083.10	1912.11
O & M cost ($\times 10^3$ \$)	0.00	3327.82	4078.81	4762.33
Energy purchased cost ($\times 10^3$ \$)	8163.28	3674.00	2646.12	2965.25
ENS cost ($\times 10^3$ \$)	10287.23	5282.39	2270.13	1953.64
Emission cost ($\times 10^3$ \$)	2789.01	1085.91	796.52	849.81
Total cost ($\times 10^3$ \$)	21239.52	14802.01	12874.67	12443.14
Savings in total cost ($\times 10^3$ \$)	0.00	6437.51	8364.85	8796.38
Total cost reduction (%)	0.00	30.31	39.38	41.42

Table 9.5 Percentage power share between RES and grid in different cases.

Parameters	Case A	Case B	Case C	Case D
Wind generation	------	48.02%	53.85%	53.92%
Solar generation	------	13.00%	19.90%	15.93%
Imported from grid	100%	38.98%	26.24%	30.14%

Table 9.5 shows the percentage of power share of RET and grid to meet the load demand of the distribution system under different cases. With installation of RET, the power share from grid has been reduced from 100% to 38.98% as seen in case B. Further, it has been reduced to 26.24% with the installation of ESS units as seen in case C. This shows that installation of ESS units encourages the RET penetration. With the deployment of the CVR scheme, the power share from grid has been reduced to 30.14% as seen in case D. It can also be observed that the share of wind generation is more compared to PV generation, which is due to availability of wind over a day.

Table 9.6 summarizes the optimal location and size of installed PV-based DGs, wind-based DGs, and ESSs in different cases in initial year. As the problem is formulated for a five-year planning horizon, the aggregated capacities of installed DGs, BESs, and SOPs have been depicted in Table 9.7. It can be seen from the table that the majority of the investments are made in the initial year. The cumulative total cost has been depicted in Table 9.8.

9.3.10.4 Effect of Voltage Profile

Figure 9.10 portrays the average voltage magnitude at each node for different cases. Figure 9.11 shows the cumulative distribution of the average voltage magnitude values for different cases. In Figures 9.11-9.12, it can be noticed that there is a significant improvement in voltages due to the integration of RET as seen in case B. Further, the voltage profile has been improved in case C, with to the installation of ESS. The voltage profile in case D is more flatt and maintained at a lower portion of the permissible limit. This is due to the operation of CVR.

9.3.10.5 Effect of Energy Losses and Consumption

Figure 9.12 and Figure 9.13 show the total energy losses and energy consumption for each year for each case, respectively. With the integration

Table 9.6 Optimal location and size of RET and ESS for different cases at initial year.

Cases	DER type	(installed location: size)
Case B	Wind (kW)	(Bus no.24: 810); (Bus no.31: 530); (Bus no.32: 210)
	Solar (kW)	(Bus no.8: 450); (Bus no.14: 370); (Bus no.18: 150); (Bus no.22: 30); (Bus no.30: 30)
Case C	Wind (kW)	(Bus no. 24: 900); (Bus no. 31: 550); (Bus no. 32: 290)
	Solar (kW)	(Bus no. 8: 540); (Bus no. 14: 450); (Bus no. 18: 320); (Bus no. 22: 190); (Bus no. 30: 120)
	ESS (kWh; kW)	(Bus no.17: 2490;620); (Bus no.33: 3090;770)
Case D	Wind (kW)	(Bus no. 24: 930); (Bus no. 31: 560); (Bus no. 32: 220)
	Solar (kW)	(Bus no. 8: 330); (Bus no. 14: 390); (Bus no. 18: 180); (Bus no. 22: 60); (Bus no. 30: 190)
	ESS (kWh;kW)	(Bus no. 17: 480;120); (Bus no. 33: 350;90)

Table 9.7 Cumulative capacities of RET and ESS for different cases year-wise.

Cases	DER type	Year 1	Year 2	Year 3	Year 4	Year 5
Case B	Wind (kW)	1550	1770	1900	2030	2160
	Solar (kW)	1020	1170	1260	1350	1450
Case C	Wind (kW)	1740	1990	2130	2280	2440
	Solar (kW)	1620	1840	1960	2090	2230
	ESS (kWh; kW)	5580; 1400	6270; 1570	6640; 1660	7030; 1760	7440; 1860
Case D	Wind (kW)	1710	1980	2100	2280	2410
	Solar (kW)	1140	1350	1550	1620	1770
	ESS (kWh; kW)	820; 210	1150; 290	1860; 470	1690; 480	2050; 510

Table 9.8 Cumulative total cost for different cases year-wise (×10³ $).

Cases	Year 1	Year 2	Year 3	Year 4	Year 5
Case A	3556.39	7658.92	11964.43	16488.28	21239.52
Case B	2540.52	5423.11	8422.97	11548.13	14802.01
Case C	2274.79	4805.57	7413.10	10102.44	12874.67
Case D	2189.90	4632.95	7155.73	9757.79	12443.14

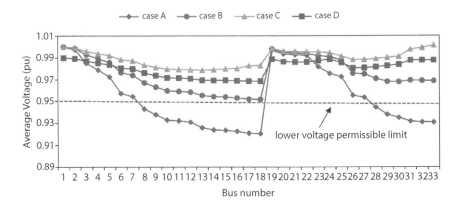

Figure 9.10 average voltage of each node for different cases.

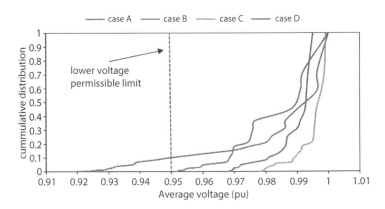

Figure 9.11 Cumulative probability distribution of nodal voltages for different cases.

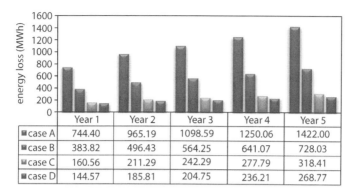

Figure 9.12 Energy losses in each year for different cases.

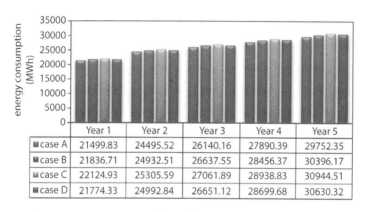

Figure 9.13 Energy consumption in each year for different cases.

of RET (i.e., case B), approximately 48.66% average reduction in losses over a five-year planning period has been achieved compared with base case (i.e., case A). Additional 29.26% average reduction in losses has been achieved with the installation of ESS (i.e., case C). Further, 3.11% more average reduction in losses has been achieved with the deployment of CVR scheme (i.e., case D). On the other hand, average energy consumption has been increased by 1.91%, 3.54% and 2.29% in case B, C and D, respectively, with respect to case A. This is due to an increase in voltage that caused an increase in energy consumption due to dependency of loads on voltage.

Similarly, Figure 9.14 shows the total energy taken from grid for each year for each case. It can be observed that total energy taken from grid has been reduced by 61.07%, 71.43% and 69.55% in case B, C and D, respectively, with respect to case A.

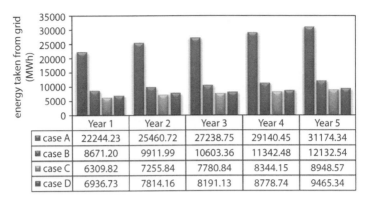

Figure 9.14 Energy taken from grid in each year for different cases.

9.3.10.6 *Effect of Energy Not Served and Carbon Emissions*

Figure 9.15 and Figure 9.16 shows the cost of CO_2 emission and cost of energy not served (ENS) respectively for each year and each case. As seen from case B through case D, it can be observed that carbon emissions have decreased due the installation of RES. On the other hand, ENS in the system has been drastically decreased in case D. This is due to the enhancement of controllability and flexibility in the system by using RET, ESS and VVC devices.

9.3.10.7 *Performance of Proposed Hybrid Optimization Solver*

In this section, the convergence performance and robustness of the hybrid optimization solver has been tested and also compared with the

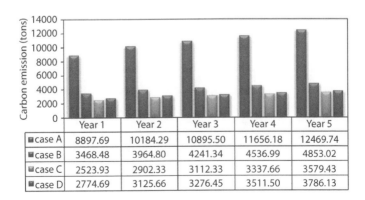

Figure 9.15 Carbon emission in each year for different cases.

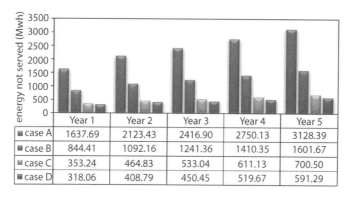

Figure 9.16 Energy not served in each year for different cases.

conventional PSO (CPSO) solver for case D in the first year of planning horizon. For this purpose, the simulation has been executed for 25 times with arbitrary initial data on Core i3 2.5GHz processor and 4 GB RAM. For both solvers, number of population and maximum iterations have been chosen as 25 and 200, respectively. Figure 9.17 shows the convergence pattern of both solvers. Further, Table 9.9 depicts the capability parameters such as the best, the worst, average and standard deviation of fitness value of both the solvers.

With the comparison of CPSO, proposed hybrid optimization solver converges faster as seen in Figure 9.17. Besides, it has achieved good capability parameters as seen in Table 9.9. It is noteworthy to mention that almost 52% of computational time has been reduced using proposed hybrid solver compared with CPSO. Further, an attempt has been made for

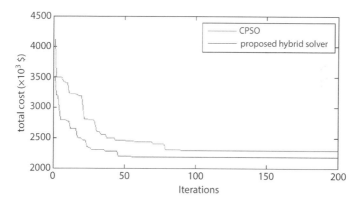

Figure 9.17 Convergence curve of different algorithms for case D.

Table 9.9 Performance of proposed hybrid solver and CPSO for case D.

Solver	Best value (×10³ $)	Worst value (×10³ $)	Average value (×10³ $)	Standard deviation (×10³ $)	Time (seconds)
PSO	2289.23	4123.5	2465.205	334.31	285.23
Proposed hybrid solver	2189.9	3512.25	2265.77	196.09	135.65

solving the present MINLP problem using the DICOPT solver in GAMS [51], but the solution process did not converge even after 1.5 hours of run; therefore it was terminated.

9.3.11 Conclusion

This chapter investigates the role of ESS in the integration of RET in an active distribution network. A two-stage coordinated optimization framework for the integrated planning model of RET and ESS incorporating a CVR scheme has been presented. Simulation results verify the effectiveness of the proposed planning model and feasibility of the proposed hybrid optimization solver. The following conclusions can be drawn from the results of the numerical simulations.

- With the installation of ESS, the share of RET can significantly be increased. Moreover, it results in a reduction in the overall cost.
- Reactive power compensation through ESS and RET devices is a potential candidate for VAR support for voltage regulation.
- Deployment of the CVR scheme can reduce the requirement of ESS capacity substantially. Moreover, it reduces network power loss, power consumption, electricity purchase cost, and CO_2 emission.
- The proposed method can improve and flatten the voltage profile without violating the permissible voltage limits.
- The proposed hybrid optimization solver saves a significant amount of computational time in solving the present MINLP problem.

It can be concluded that the proposed method is seamless and secure coordination and integration of DERs at a large scale can enhance reliability, flexibility, power quality, and resiliency while minimizing capital investment and operating costs.

References

1. Global tends in renewable energy investment, 2019, https://www.unenvironment.org/resources/report/global-trends-renewable-energy-investment-2019.
2. Ministry of New and Renewable Energy (MNRE), 2019, https://mnre.gov.in/.
3. N. Acharya, P. Mahat, and N. Mithulananthan., An analytical approach for dg allocation in primary distribution network. *International Journal of Electrical Power & Energy Systems*, vol. 28, no. 10, pp. 669-678, 2006.
4. D. Q. Hung, N. Mithulananthan, and R. Bansal., Analytical expressions for dg allocation in primary distribution networks. *IEEE Transactions on Energy Conversion*, vol. 25, no. 3, pp. 814-820, 2010.
5. D. Q. Hung and N. Mithulananthan., Multiple distributed generator placement in primary distribution networks for loss reduction. *IEEE Transactions on Industrial Electronics*, vol. 60, no. 4, pp. 1700-1708, 2013.
6. C. Wang and M. H. Nehrir., Analytical approaches for optimal placement of distributed generation sources in power systems. *IEEE Transactions on Power Systems*, vol. 19, no. 4, pp. 2068-2076, 2004.
7. G. Celli, E. Ghiani, S. Mocci, and F. Pilo., A multi-objective evolutionary algorithm for the sizing and siting of distributed generation. *IEEE Transactions on Power Systems*, vol. 20, no. 2, pp. 750-757, 2005.
8. S.-H. Lee and J.-W. Park., Selection of optimal location and size of multiple distributed generations by using kalman filter algorithm. *IEEE Transactions on Power Systems*, vol. 24, no. 3, pp. 1393-1400, 2009.
9. S. H. Lee and J.-W. Park., Optimal placement and sizing of multiple DGs in a practical distribution system by considering power loss. *IEEE Transactions on Industry Applications*, vol. 49, no. 5, pp. 2262-2270, 2013.
10. M. F. Shaaban, Y. M. Atwa, and E. F. El-Saadany., Dg allocation for benefit maximization in distribution networks. *IEEE Transactions on Power Systems*, vol. 28, no. 2, pp. 639-649, 2013.
11. S. N. G. Naik, D. K. Khatod, and M. P. Sharma., Analytical approach for optimal siting and sizing of distributed generation in radial distribution networks. *IET Generation, Transmission & Distribution*, vol. 9, no. 3, pp. 209-220, 2014.
12. A. Ameli, S. Bahrami, F. Khazaeli, and M.-R. Haghifam., A multi objective particle swarm optimization for sizing and placement of DGs from DG

owner's and distribution company's viewpoints. *IEEE Transactions on Power Delivery*, vol. 29, no. 4, pp.1831-1840, 2014.
13. S. Elsaiah, M. Benidris, and J. Mitra., Analytical approach for placement and sizing of distributed generation on distribution systems. *IET Generation, Transmission & Distribution*, vol. 8, no. 6, pp. 1039-1049, 2014.
14. V. Murthy and A. Kumar., Comparison of optimal DG allocation methods in radial distribution systems based on sensitivity approaches. *International Journal of Electrical Power & Energy Systems*, vol. 53, pp. 450-467, 2013.
15. H. L. Willis., Analytical methods and rules of thumb for modelling DG-distribution interaction. in *Power Engineering Society Summer Meeting, 2000. IEEE, vol. 3. IEEE*, 2000, pp. 1643-1644.
16. K. M. Muttaqi, A. D. Le, M. Negnevitsky, and G. Ledwich., An algebraic approach for determination of DG parameters to support voltage problems in radial distribution networks. *IEEE Transactions on Smart Grid*, vol. 5, no. 3, pp. 1351-1360, 2014. 178.
17. D. Kumar, S. Samantaray, I. Kamwa, and N. Sahoo., Reliability-constrained based optimal placement and sizing of multiple distributed generators in power distribution network using cat swarm optimization. *Electric Power Components and Systems*, vol. 42, no. 2, pp. 149-164, 2014.
18. M. Esmaili., Placement of minimum distributed generation units observing power losses and voltage stability with network constraints. *IET Generation, Transmission & Distribution,* vol. 7, no. 8, pp. 813-821, 2013.
19. T. Shukla, S. Singh, V. Srinivasarao, and K. Naik., Optimal sizing of distributed generation placed on radial distribution systems. *Electric Power Components and Systems*, vol. 38, no. 3, pp. 260-274, 2010.
20. W. S. Tan, M. Y. Hassan, H. A. Rahman, M. P. Abdullah, and F. Hussin., Multi-distributed generation planning using hybrid particle swarm optimisation gravitational search algorithm including voltage rise issue. *IET Generation, Transmission & Distribution*, vol. 7, no. 9, pp. 929-942, 2013.
21. P. S. Georgilakis and N. D. Hatziargyriou., Optimal distributed generation placement in power distribution networks: models, methods, and future research. *IEEE Transactions on Power Systems*, vol. 28, no. 3, pp. 3420-3428, 2013.
22. N. Jain, S. Singh, and S. Srivastava., A generalized approach for dg planning and viability analysis under market scenario. *IEEE Transactions on Industrial Electronics*, vol. 60, no. 11, pp. 5075-5085, 2013.
23. R. Li, W. Wang, X. Wu, F. Tang, and Z. Chen., Cooperative planning model of renewable energy sources and energy storage units in active distribution systems: A bi-level model and pareto analysis. *Energy*, vol. 168, pp. 30-42, 2019.
24. J. M. Home-Ortiz, M. Pourakbari-Kasmaei, M. Lehtonen, and J. R. S. Mantovani., Optimal location-allocation of storage devices and renewable-based dg in distribution systems. *Electric Power Systems Research*, vol. 172, pp. 11-21, 2019.

25. Y. Li, B. Feng, G. Li, J. Qi, D. Zhao, and Y. Mu., Optimal distributed generation planning in active distribution networks considering integration of energy storage. *Applied Energy*, vol. 210, pp. 1073-1081, 2018.
26. M. R. Jannesar, A. Sedighi, M. Savaghebi, and J. M. Guerrero., Optimal placement, sizing, and daily charge/discharge of battery energy storage in low voltage distribution network with high photovoltaic penetration. *Applied Energy*, vol. 226, pp. 957-966, 2018.
27. V. A. Evangelopoulos and P. S. Georgilakis., Optimal distributed generation placement under uncertainties based on point estimate method embedded genetic algorithm. *IET Generation, Transmission & Distribution*, vol. 8, no. 3, pp. 389-400, 2013.
28. Y. Zhang, S. Ren, Z. Y. Dong, Y. Xu, K. Meng, and Y. Zheng., Optimal placement of battery energy storage in distribution networks considering conservation voltage reduction and stochastic load composition. *IET Generation, Transmission & Distribution*, vol. 11, no. 15, pp. 3862-3870, 2017.
29. L. Bai, T. Jiang, F. Li, H. Chen, and X. Li., Distributed energy storage planning in soft open point based active distribution networks incorporating network reconfiguration and dg reactive power capability. *Applied Energy*, vol. 210, pp. 1082-1091, 2018.
30. L. Zhang, C. Shen, Y. Chen, S. Huang, and W. Tang., Coordinated allocation of distributed generation, capacitor banks and soft open points in active distribution networks considering dispatching results. *Applied Energy*, vol. 231, pp. 1122-1131, 2018.
31. A. Ali, D. Raisz, and K. Mahmoud., Optimal oversizing of utility-owned renewable dg inverter for voltage rise prevention in mv distribution systems. *International Journal of Electrical Power & Energy Systems*, vol. 105, pp. 500-513, 2019.
32. S. S. AlKaabi, V. Khadkikar, and H. Zeineldin., Incorporating pv inverter control schemes for planning active distribution networks. *IEEE Transactions on Sustainable Energy*, vol. 6, no. 4, pp. 1224-1233, 2015.
33. M. Nick, R. Cherkaoui, and M. Paolone., Optimal planning of distributed energy storage systems in active distribution networks embedding grid recon_guration. *IEEE Transactions on Power Systems*, vol. 33, no. 2, pp. 1577-1590, 2017.
34. S. Zhang, H. Cheng, D. Wang, L. Zhang, F. Li, and L. Yao., Distributed generation planning in active distribution network considering demand side management and network reconfiguration. *Applied Energy*, vol. 228, pp. 1921-1936, 2018.
35. M. Gheydi, A. Nouri, and N. Ghadimi., Planning in microgrids with conservation of voltage reduction. *IEEE Systems Journal*, vol. 12, no. 3, pp. 2782-2790, 2016.
36. Y. Zhang, Y. Xu, H. Yang, and Z. Y. Dong., Voltage regulation-oriented co-planning of distributed generation and battery storage in active distribution

networks. *International Journal of Electrical Power & Energy Systems*, vol. 105, pp. 79-88, 2019.
37. M.Wu, L. Kou, X. Hou, Y. Ji, B. Xu, and H. Gao., A bi-level robust planning model for active distribution networks considering uncertainties of renewable energies. *International Journal of Electrical Power & Energy Systems*, vol. 105, pp. 814-822, 2019.
38. D. Singh, D. Singh, and K. Verma., Multiobjective optimization for DG planning with load models. *IEEE Transactions on Power Systems,* vol. 24, no. 1, pp. 427-436, 2009.
39. M. F. Shaaban, Y. M. Atwa, and E. F. El-Saadany., DG allocation for benefit maximization in distribution networks. *IEEE Transactions on Power Systems*, vol. 28, no. 2, pp. 639-649, 2012.
40. M. Moradi and M. Abedini., A combination of genetic algorithm and particle swarm optimization for optimal DG location and sizing in distribution systems. *International Journal of Electrical Power & Energy Systems*, vol. 34, no. 1, pp. 66-74, 2012.
41. S. Montoya-Bueno, J. I. Muoz, and J. Contreras., A stochastic investment model for renewable generation in distribution systems. *IEEE Transactions on Sustainable Energy*, vol. 6, no. 4, pp. 1466-1474, 2015.
42. O. D. Melgar-Dominguez, M. Pourakbari-Kasmaei, and J. R. S. Mantovani., Adaptive robust short-term planning of electrical distribution systems considering siting and sizing of renewable energy based dg units. *IEEE Transactions on Sustainable Energy*, vol. 10, no. 1, pp. 158-169, 2018.
43. E. Samani and F. Aminifar., Tri-level robust investment planning of DERs in distribution networks with ac constraints. *IEEE Transactions on Power Systems*, 2019.
44. O. Gandhi, C. D. Rodríguez-Gallegos, W. Zhang, D. Srinivasan, and T. Reindl., Economic and technical analysis of reactive power provision from distributed energy resources in microgrids. *Applied Energy*, vol. 210, pp. 827-841, 2018.
45. V. Murty and A. Kumar., Mesh distribution system analysis in presence of distributed generation with time varying load model. *International Journal of Electrical Power & Energy Systems*, vol. 62, pp. 836-854, 2014.
46. S. F. Santos, D. Z. Fitiwi, M. R. Cruz, C. M. Cabrita, and J. P. Catalão., Impacts of optimal energy storage deployment and network reconfiguration on renewable integration level in distribution systems. *Applied Energy*, vol. 185, pp. 44-55, 2017.
47. M. Diaz-Aguiló, J. Sandraz, R. Macwan, F. De Leon, D. Czarkowski, C. Comack, and D. Wang, "Field-validated load model for the analysis of cvr in distribution secondary networks: Energy conservation," *IEEE Transactions on Power Delivery,* vol. 28, no. 4, pp. 2428–2436, 2013.
48. F. Scarlatache, G. Grigoras, G. Chicco, and G. Cârtină., Using kmeans clustering method in determination of the optimal placement of distributed generation sources in electrical distribution systems. In *2012 13th International*

Conference on Optimization of Electrical and Electronic Equipment (OPTIM). IEEE, 2012, pp. 953–958.
49. P. Vijay Babu, and S. P. Singh., Optimal coordination of PV smart inverter and traditional volt-VAR control devices for energy cost savings and voltage regulation." *International Transactions on Electrical Energy Systems*, vol. 29, no. 7 (2019): e12042.
50. P. V. Babu and S. Singh., Optimal placement of dg in distribution network for power loss minimization using nlp & pls technique. *Energy Procedia*, vol. 90, pp. 441–454, 2016.
51. I. E. Grossmann, J. Viswanathan, A. Vecchietti, R. Raman, E. Kalvelagen *et al.*, GAMS/DICOPT: A discrete continuous optimization package. GAMS Corporation Inc, vol. 37, p. 55, 2002.
52. M. E. Baran and F. F. Wu., Network reconfiguration in distribution systems for loss reduction and load balancing. *IEEE Transactions on Power Delivery*, vol. 4, no. 2, pp. 1401–1407, 1989.
53. Cruz, M. R., Fitiwi, D. Z., Santos, S. F., Mariano, S. J., & Catalão, J. P., Multi-Flexibility Option Integration to Cope With Large-Scale Integration of Renewables. *IEEE Transactions on Sustainable Energy*, vol. 11(1), pp. 48-60, 2018.

10

Inclusion of Energy Storage System with Renewable Energy Resources in Distribution Networks

Rayees Ahmad Thokar[1]*, Vipin Chandra Pandey[1], Nikhil Gupta[1], K. R. Niazi[1], Anil Swarnkar[1], Pradeep Singh[2] and N. K. Meena[3]

[1]Department of Electrical Engineering, Malaviya National Institute of Technology, Rajasthan, India
[2]Indian Institute of Technology, Delhi, India
[3]School of Engineering and Applied Science, Aston University, Birmingham, United Kingdom

Abstract

The rapid growth of renewable energy resources in modern distribution networks results in the spilling of energy due to the limited hosting capacity of these networks, intermittent and variable nature of renewable generation, violation of system constraints, reduced network efficiency, and improper utilization of resources. Energy storage system has emerged as a key technology element in improving the operational efficiency, reliability, security, flexibility and performance of renewable energy technology integrated distribution networks. The storage technology is regarded as one of the potential candidates in facilitating increased penetration of renewable energy technologies thereby increasing the renewable generation hosting capacity of the distribution networks. However, the benefits provided by these storage technologies to the distribution network operators, energy storage owners, and end-customers are still questionable and full of queries. In this chapter, a thorough summary of the ongoing studies on the deployment (location, capacity and daily dispatch) of energy storage system technologies in the distribution networks is presented. Further, the chapter also gives a distinctive understanding of the various challenges, issues, recommendations, and benefits of

---

*Corresponding author: raishameed427@gmail.com; 2015ree9542@mnit.ac.in

Sandeep Dhundhara and Yajvender Pal Verma (eds.) Energy Storage for Modern Power System Operations, (281–328) © 2021 Scrivener Publishing LLC

incorporating energy storage technologies into the modern distribution networks thus offering a benchmark for carrying out further research in this area. Finally, a case study is performed highlighting the significance of simultaneous allocation of battery energy storage technology and renewable energy resources in a 33-bus test distribution network.

Keywords: Energy storage technology, renewable energy resources (RERs), distribution network, ESS allocation, battery energy storage systems (BESSs), operation-planning, integration of energy storage and renewable energy technologies, joint optimization model

10.1 Introduction

In recent years, with the rising climatic threats, increasing carbon footprints, transformation of energy policy and increasing energy crisis, national governments around the world have been forced to impose integration of renewable energy technologies on a large scale and considerable changes on how to operate the modern distribution networks (DNs) so as to provide reliable, efficient and secure power supply. Further, a promising development of Smart Distribution Network (SDN) concepts within vertically integrated systems can be made possible with the help of optimal Distributed Renewable Energy Resources (DRERs) integration.

Considering these realities, nowadays power system researchers and national governments around the globe are enthusiastically encouraging the replacement of traditional fossil fuel–based energy with green and clean renewable-based energy technologies to promote the sustainable development of mankind and society in every walk of life. Therefore, the modern power system has steered in a new energy revolution and DRERs such as Solar Photovoltaic (SPV) and Wind Turbine (WT) power generators are being widely used on the DN side. However, the renewable energy technology integration in modern distribution systems has both pros and cons. The various benefits provided by the optimal integration of DRERs in contemporary distribution systems include enhancement in reliability, stability and power quality, reduction in carbon footprints, power/energy loss minimization, minimization of node voltage deviation, peak load shaving, system investment deferral, and so on. While the DRERs bring green and clean power to the grid, the flexibility caused by its volatility and intermittence also puts new challenges on the operation and planning of the DNs [1–3]. Further, their non-optimal allocation may bring a chain of system operational issues which may include fault current rising, worsening of load flatness index, duck curve problem, voltage regulation issues, energy losses and many more. Also, the rapid growth of integration of renewables technologies in modern distribution networks results in the spilling of energy due to the limited hosting capacity of these networks, violation of system and DRER constraints, reduced network efficiency, and improper utilization of resources. However, these negative impacts may be mitigated by optimal placement, sizing and/or charge/discharge dispatch of energy storage system [4].

For decades different types of energy storage system (ESS) technologies have played a role in providing arbitrage or contingency services in the electrical power industry. Therefore, these storage technologies do not represent any new energy subjects nor do they provide any new concepts regarding the operation of modern distribution networks. Nevertheless, in case of large energy storage units like pumped hydro energy storage and compressed air energy storage the number, placement, and accordingly capacity, in terms of power and energy, are limited due to the prevailing geographical conditions and exhaustive infrastructure requirements. In contrast, the smaller energy storage units, i.e., battery energy storage systems (BESSs), are unrestricted from the above-mentioned limitations; however, they necessitate higher initial investment cost as compared to the larger ones. Due to this limitation, the BESSs have been used previously only for specific applications like backup supply purpose.

Recently, the ESS has gained sufficient reputation in the electrical power industry due to its ability to improve the stability and flexibility of the modern power system, provide ride through capability during loss of generation, perform energy and price arbitrage and also mitigate the effect of intermittency and volatility caused by the RER-based Distributed Energy Resources (DERs) [5, 6]. ESSs are the among key technology elements in renewable energy technology integrated distribution systems. These elements are regarded as one of the most viable solutions for facilitating increased penetration of renewable energy technologies [7, 8]. The optimal deployment of these dispatchable storage technologies can provide multiple benefits for Distribution Network Operators (DNOs), DER owners, and end-consumers through greater reliability, improved power quality, proliferation of arbitrage benefits, and overall reduced energy costs. The BESSs not only inject/absorb energy into/from the system during stringent system conditions but can also provide inertial response (digital inertia) for enhanced system reliability and flexibility much more efficiently at a lower cost and with substantially reduced emissions than a much larger quantity of conventional generation (analogue inertia). These days, the energy storage deployment also encourages several grid ancillary services called as distributed ancillary services [9]. Furthermore, the effective management of these storage technologies generates profit from energy price arbitrage, peak-load shaving, load-shifting, demand response programs, and improved operational efficiency of the modern power system. To sum up, one can say, despite being costly, having a shorter lifespan and complex control, ESS offers a techno-economically feasible solution for many issues and challenges encountered in SDNs. In short, SDNs are represented by storage technologies in numerous ways.

The chapter is planned as follows. Section 10.2 presents the energy storage allocation and its methods. The multiple applications of these storage technologies in modern power systems are presented in Section 10.3. Section 10.4 reviews several types of storage technologies for smart and sustainable operation of modern power networks. In Section 10.5, a case study is presented. The case study discusses the simultaneous allocation of multiple BESS and SPV units in distribution network by framing a joint optimization model. Further, the section also presents results and conclusions of the case study followed by directions for future research and recommendations in Section 10.6.

10.2 Optimal Allocation of ESSs in Modern Distribution Networks

Despite being sufficiently advantageous, the optimal operational planning of an energy storage system (ESS) i.e., the optimal location, sizing and daily dispatch together with existing DRERs is highly essential to enhance the system operational efficiency and performance. It is not an economic option to install large ESS at every bus, especially in a large DN, as the non-optimal deployment of ESS may further burden the utilities and other stakeholders with higher investment cost and may drastically reduce the arbitrage benefits. Also, the optimal and effective integration of storage technologies especially BESS is a challenging task unlike DRER integration. It is limited by timely availability of state-of-charge (SOC), its associated constraints, charging-discharging decisions, number of life cycles, and dispatch control [10, 11]. However, the well-coordinated and effective operational management of ESSs in SDNs can systematically optimize the network operations, make DRERs act as dispatchable sources, help in improving the load flatness index, and so on. This section presents a comprehensive review of different roles that BESSs can play in the modern power networks and the methodologies, strategies and optimization techniques employed to optimally allocate ESSs in DNs.

10.2.1 ESS Allocation (Siting and Sizing)

As mentioned above, the strategically deployed and optimally operated ESS units can provide numerous benefits to DSOs, ESS owners and other stakeholders. Contrarily, the non-optimal allocation of ESSs in DNs may provide counter-productive results [12]. Depending on several categories of ESS technologies, diverse system requirements and the applications for which these technologies are being used, the optimal placement and sizing problem of ESSs in DNs lack a preferred or unified solution [13]. The planning and operation of ESSs cannot be perceived separately because these are embedded. Therefore, the operational planning problem of storage technologies represents a complex nested optimization problem which may require a well-coordinated and well-tailored multi-level strategy and optimization framework.

A significant amount of literary work is presented in the area of ESS allocation in terms of optimization and selection of placement, sizing and type of ESS units in modern distribution networks [14–17]. In reviewed literature, several planning and operation strategies for the optimal ESS allocation have been suggested in the modern distribution networks. A number of optimization models are proposed to optimally allocate ESSs in renewable energy technology integrated DNs [18–20]. To mitigate the system cost and to enhance the system reliability and performance, an optimization model for energy storage allocation is proposed in [18] while considering the uncertainty of wind power generation. Optimal location and capacity of battery storage in distribution network are determined in [19] by proposing a chance constrained optimization model while employing Monte-Carlo simulation to handle the uncertainty and variability in wind generation and load, respectively. In [20], the authors proposed a negative sensitivity index–based method for optimal storage deployment in distribution networks while mitigating voltage deviation and system power losses. The optimal integration of BESS along with daily charge/discharge dispatch is determined in [4] by utilizing a combination of evolutionary algorithm and linear programming method while considering optimization of a cost function. Li et al. [21] proposed an interesting two-stage optimization problem for optimally integrating energy storage system and distributed generation in DNs. As per [22], diverse storage technologies can in general be recognized by numerous characteristics which may include location, power and energy capacity, frequency response, control, implementation requirement, time response, and so on. However, in most of the literature ESSs are mainly chosen and optimized depending on their placement, power rating and energy capacity in the electrical power network. The power rating and energy capacity are discrete variables, although, at times these are treated as continuous variables and several optimization approaches are utilized to solve them. Exhaustive search methods [23, 24] have been found to be of greater use while defining these two characteristics as discrete variables. Further, charging and discharging efficiencies (calculated as square root of the round trip efficiency) or charging and discharging power as a constraint must be accounted for when these ESS characteristics are modeled.

The optimal integration of ESS while considering location, sizing, daily charge/discharge dispatch and determination of instantaneous SOC using several estimation algorithms simultaneously turns out to be a complex, mixed-integer, non-linear and multi-constraint two-level optimization problem. In literature, numerous methods have been proposed for ESS allocation in modern distribution networks so as to gain the optimum planning and operational solutions. Depending on the methodologies, strategies and optimization techniques employed these methods are grouped into four main categories which can be defined as analytical methods, mathematical programming, exhaustive search and heuristic methods. A brief literature survey considering these methods is presented in the section below.

10.2.2 ESS Allocation Methods

A. *Analytical Methods*
Generally, analytical methods (AM) make use of a predefined system and operational constraints instead of specific system data. Recently, these methods are employed to make DRERs act as dispatchable sources by optimally sizing ESSs in distribution networks. Numerous authors have presented the role of ESSs as a viable solution to absorb the energy from renewable technologies [25, 26]. Besides recognizing the potential benefits of optimal ESS placement in power system these methods are normally restricted to obtaining energy price arbitrage benefits. In literature, various analytical methods based on historical load demand curves or numerical data explorations are presented in [27–34]. These methods do not take into account market-driven operation or network constraints. On the contrary, the optimal energy storage sizing is determined with the existing storage placement in [35] while considering energy price arbitrage benefits in the electrical power network. Analytical methods based on statistical wind distribution and a combination of wind power forecast error distribution and energy not supplied are proposed in [28] and [31], respectively, for

determining the power rating and energy capacity of storage technology. In [29], charging and discharging power and sizing of ESS are obtained depending on the system load demand and availability and unavailability of solar power generation. Reliability-based optimal sizing of ESS to be used as a backup source is proposed in [30], whereas in [36] desired peak shaving and real system demand are used to calculate battery storage system capacity. The implementation and execution of analytical methods are easy and fast, respectively, but their results may be impractical due to the presence of certain simplified assumptions. For this reason, these methods may sometimes fail to solve the complex, real-life engineering optimization problems.

B. *Mathematical Programming*
Mathematical programming constitutes diverse numerical methods that are capable of determining optimum solution of designed problem frameworks. Among these methods, linear programming [37–39] is the most competent, well-organized and effective method. This method always has the potential of finding a unique global optimal solution. Further, linear programming is beneficial in maintaining the linearity of the designed problem when the power network shows irrational behaviour towards linearity. Several literary works are presented to optimally allocate ESSs in distribution networks by employing mixed-integer linear programming method. In [37] the authors analyzed the impact of wind penetration and operation policy of diesel generator on ESS sizing and network energy cost. The case study concludes that the ESS allocation is justified only for distribution networks integrated with medium to high level of renewable energy technologies. The optimal sizing and siting of ESS are determined in [38] while minimizing spilled wind energy and network losses, respectively. The authors in [39] demonstrated that the energy storage allocation can deliver potential benefits across an electrical power network while supporting real-time network balancing, investment deferral, network operational cost minimization and network congestion management.

In literature, several other mathematical programing methods for solving energy storage integration problems in DNs may include second-order cone programming [40, 41], dynamic programing [42], stochastic programing [43], and many more. In [40] a multi-objective optimization is solved while minimizing cost of energy storage investment and network operational cost. In this work, the authors concluded that a limited number of ESS units placed optimally can effectively control the operation of distribution network integrated with numerous numbers of DG units. Feasible investment case for optimal battery storage system capacity and placement is realized in DNs in [41] while achieving arbitrage benefits and mitigation of network losses for the operator. However, the authors did not succeed in finding a suitable business case for investment on BESSs. The role of battery storage technology as a potential candidate in contract capacity saving is evaluated in [42] while minimizing grid contract and storage investment cost. In [43], the optimal energy storage sizing, placement and daily dispatch is determined while minimizing the cost of storage investment and network operation simultaneously. As discussed, most of the mathematical programming methods guarantee the optimal global solution of the problem; however, these are not realizable for large-scale systems. This is also the case with dynamic programing. These methods also suffer from the difficulty of providing only a single optimal solution for a problem, which may not be feasible for ESS deployment due to strategic location, political reasons, etc.

C. *Exhaustive Search*

Exhaustive search method is highly computationally demanding even for the optimization problems which are moderate in size. However, it is promising in finding the global optimum solution in a restricted problem search space. The optimal energy capacity and power rating of ESS are obtained using exhaustive search in numerous literary works; whereas, in [44] only storage system placement problem is evaluated independently from capacity evaluation. Further, due to the complex combinatorial nature of the ESS allocation problem, practically exhaustive search methods are incompetent to solve the optimal siting and sizing problem of storage systems simultaneously.

In literature, a diversity of ESS allocation problems has been solved by employing the exhaustive search method. In these research efforts, numerous combinations of exhaustive search with simulation techniques are put into action to achieve the optimal energy capacity and power rating of ESS. The objective functions considered in these research works include reduction of network cost and reliability enhancement [45], maximization of fuel saving [23], minimizing frequency deviation, minimizing charge/discharge dispatch cycles of battery storage systems [24], minimization of ESS investment cost and imbalance penalties [46], minimizing net present value of ESS installation [47], etc. The different frameworks and simulation techniques utilized while optimizing these objectives include dynamic programming [23], mathematical programming [24], Monte-Carlo simulation [43], genetic algorithm [47], and so on. Besides guaranteeing a promising global optimal solution, exhaustive search also suffers from the impracticality for solving complex power system optimization problems like mathematical programming methods.

D. Heuristic Methods

Heuristic methods represent a class of algorithms that work on the basis of experience and knowledge concerning a particular problem. Recently, these methods have become a substitute term for evolutionary or swarm intelligence–based computational techniques that are inspired by the concept of natural selection or evolutionary process. These AI-based, nature-inspired methods have the capability of exploring solution search space in a smart and effective manner. The heuristic methods are highly feasible for handling complex power system optimization problems; however, they do not show promise in finding a global optimal solution. Besides being proven to be robust in nature their outcomes are practically highly feasible. Alhough these methods are usually highly computationally demanding, this limitation does not seem to be necessarily crucial in optimally integrating ESSs in modern power systems.

In literature, a diversity of heuristic methods has been suggested to solve the siting and sizing problem of BESSs simultaneously. In this optimization problem, usually the complex combinatorial part is solved by the heuristic methods and the solution obtained thereafter is further evaluated by some other methods (generally used for solving optimal power flow (OPF) or unit commitment (UC)). Some of the powerful and popularly known heuristic techniques adopted for ESS allocation may include genetic algorithm (GA) [48, 49], particle swarm optimization (PSO) [50–53], artificial bee colony (ABC) [54], bat algorithm (BA) [55], African buffalo optimization (ABO) [10], and so on. In [48], optimal sizing and placement of multiple DERs including ESS are determined by employing a combination of GA and sequential quadratic programming technique (SQP). In this work, GA is used to calculate optimal capacity and placement of DER units whereas SQP is used to solve OPF while minimizing cost of network operation. The authors in [49] proposed an optimization framework comprising GA and probabilistic OPF to optimally integrate energy storage units in power network while maximizing energy price arbitrage and minimizing ESS investment cost. The authors concluded that storage investment becomes economically feasible due to adequate arbitrage benefits obtained only for the higher levels of wind power integration in the distribution network. A modified PSO with improved local search ability is proposed in [50] to optimally integrate DG and storage units in DN. In [52], the authors proposed a fuzzy-based PSO to solve the optimal ESS allocation problem while minimizing network losses and error in load curve prediction. Further, to evaluate DG and ESS performance in [50] and [52] a simple rule-based method is employed instead of an OPF while calculating cost of network operation. The optimal placement and sizing of battery switching station units are determined in [54] by utilizing an ABC technique while minimizing network losses and maximizing battery storage utilization. In [55], the optimal capacity of DG-BESS system is determined with predefined battery energy storage placement by using a new improved BA solution technique while minimizing network operation cost and storage investment cost.

Therefore, AI-based nature-inspired algorithms which are mostly swarm intelligence–based [50–55] or evolutionary-based algorithms [48, 49] have shown their promising performance and thus have become popular and widely accepted for solving energy storage allocation problems in a well-organized and competent way. Furthermore, the issues, challenges and benefits of different ESS allocation methods are summarized in Table 10.1.

10.3 Applications of ESS in Modern Distribution Networks

The potential benefits of energy storage technologies in modern power systems, presently and in the near future, can be diverse and numerous. These benefits are mainly based on the various services these technologies provide at multiple positions in the modern energy networks. In short, energy storage technologies possess the potential to be used all round the electrical power industry, which means ESS deployment could be realized across the supply, transmission and distribution and demand (end-consumer) side of the future energy network. Moreover, across supply and demand portions of the future energy networks, they are capable of providing sufficient infrastructural support to enhance the integration and penetration of renewable energy technologies. In a wider perspective, storage technologies can act as feasible facilities for network operator, which is evident from the reviewed literature, with highly intermittent and variable supply and demand side [56]. At present, most of the literary works are focused on the exploitation of ESS deployment benefits by a single stakeholder; however, they can provide numerous benefits to multiple network stakeholders. The role of these technologies, encouraged by the purely economic motives, in offering benefits to DNOs, ESS owners and other stakeholders, is to uphold the system within certain specified system and operation constraints and factors [14]. Considering the exploitation of ESS deployment benefits by multiple stakeholders, the contribution of ESS is acknowledged either on generation and distribution side or at the end-consumer side. In this regard, the reviewed literature on ESS applications and benefits in modern power system can be categorized as shown below.

Table 10.1 Several issues, challenges and benefits of ESS allocation methods.

ESS allocation methods		Issues & challenges	Benefits
Conventional approach	**Analytical methods:** 2/3 thumb rule Exact loss formula Convex relaxation method Loss sensitivity factor–based method Cost-benefit analysis–based method Benders decomposition method Interior point algorithm, etc.	Indicative results due to simplified assumptions Ineffective for solving complex engineering optimization problems	Easy to implement Fast to execute
	Exhaustive search	Not feasible for large-scale systems Faces difficulty in solving joint optimization problems independently Suffers from curse of dimensionality Highly computationally demanding	Guarantees global optimum
Mathematical programming	Linear programming Non-linear programming Dynamic programming	Not feasible for large-scale systems and multi-objective optimization problems	Computationally fast and efficient Promises global optimal solution

(Continued)

Table 10.1 Several issues, challenges and benefits of ESS allocation methods. (*Continued*)

ESS allocation methods		Issues & challenges	Benefits
	Chance constrained-programming Mixed-integer programming Mixed-integer non-linear programming Stochastic programming Second order cone programming Sequential quadratic programming Monte Carlo simulation, etc.	Need very accurate modelling for problem formulation Provides only single optimal output	
Heuristic approach	**Swarm intelligence–based algorithms:** Standard variants (PSO, ABO, MSO, ACO, ABC, BA, etc.) Improved versions (IPSO, GPSO, CMSO, MABO, IBA, etc.)	Computationally intense Local trapping and poor performance while seeking global optimum solutions (standard variants)	Robust in nature Highly feasible output Fast convergence to seek global optimum (improved versions) Highly efficient and feasible for solving complex optimization problems
	Evolutionary-based algorithms: Standard variants (GA, DE, etc.) Improved versions (NSGA-II, DNPL based GA, IGA, hybrid DE, etc.)	Don't promise a global optimum	Provides optimal or near optimal solution Very effective for solving multiobjective and joint optimization problems (mostly improved versions)

MSO (moth search optimization); ACO (ant colony optimization); DE (differential evolution); IPSO (improved PSO); GPSO (gradient PSO); CMSO (corrected MSO); IBA (improved BA); NSGA-II (non-dominated sorting genetic algorithm-II); DNPL (dynamic node priority list); IGA (improved GA).

10.3.1 ESS Applications at the Generation and Distribution Side

The energy storage technologies can contribute in making the non-dispatchable energy sources act as dispatchable, thereby smoothing the power output of these intermittent resources [57–59]. The fluctuating output of these resources can affect the frequency and voltage of distribution network which can lead to degraded power quality [57]. The influence of this fluctuating output on the power network can be mitigated by quickly charging ESS during excess renewable generation and discharging during low renewable generation [58, 60, 61]. Therefore, the key attribute of the energy storage technology in making the renewable generation output dispatchable and smooth is its rapid ramping capability [34, 35, 58].

The ESS can also provide energy arbitrage benefits by storing energy from dispatchable and non-dispatchable sources during low off-peak or low-price periods and consequently selling it during on-peak or high-price periods [36, 62]. This time shifting of power generation is possible only if the storage technology is capable of storing a large quantity of energy for a sufficient amount of time, usually ranging from a few hours to a few days [58]. The ESS can also be employed for long-term time shifting of generation output, usually ranging from a few days to a few months, in the form of seasonal storage facility [57, 59]. They can play a crucial role in regulating the voltage profile and preventing voltage collapse of distribution networks by injecting and managing the reactive power flow in the network, respectively [58, 59]. Moreover, during any sudden disturbance on the generation or distribution side the storage technology can supply load to reduce voltage sags and swells thus supporting the voltage profile of the network [63].

In addition to the above-discussed roles, storage technologies can help in minimizing network investment and operational cost [64–66], forecast error [31, 67], wind curtailment [49, 68, 69], carbon footprints [70], energy cost [37, 51, 71], and so on. They can also help in mitigating spillage of energy or reverse feed resulting from rapid DG integration [72], in providing spinning reserve [58, 63], in peak shaving [57], and in avoiding system violations [73]. In case of any power failure or power blackout ESS can act as feasible technology for supporting black start [58, 63].

10.3.2 ESS Applications at the End-Consumer Side

Due to relatively higher initial investment cost of energy storage technologies previously they have been primarily deployed for only specialized applications such as providing backup supply [30, 33]. Instead of conventional generators they can provide backup power supply for short duration outages at less expense [74]. The applications of ESS at consumer-end side are more related to fulfilling power and energy demand [13] and improving network reliability [30, 33]. Moreover, at the consumer-end ESSs are crucial in enhancing the ride-through capability of industrial load and residential appliances by providing an uninterrupted power supply [13, 59]. The ESS technologies can also help in providing capacity contract savings or peak reduction [57], proliferation of arbitrage benefits [36, 75], demand response and smart and sustainable home [76, 77].

For better and easy understanding, the above-discussed ESS applications considering multiple power system stakeholders are further summarized in Table 10.2.

In addition to the above discussion, the suitability of a particular technology for an individual application can be broadly evaluated in terms of its technical potential. For energy storage technology, power and energy capacity, discharge duration and time to response provide a worthy and early indicator on suitability [78]. Depending on these key attributes some distinctive energy storage technology applications at both the supply and demand side are enumerated as presented in Table 10.3.

10.4 Different Types of ESS Technologies Employed for Sustainable Operation of Power Networks

The numerous types of energy storage technologies which can be implemented for sustainable and smart operation of modern power networks may include compressed air energy storage, pumped hydro energy storage, fuel cells, flywheel energy storage, super capacitor, superconducting magnetic energy storage, and battery energy storage systems. A comparison of several key technical features of these technologies is presented in Table 10.4 [57, 60, 79]. The data given in this table is an example of representative features from the reviewed literature but not a comprehensive list of features. Most of these parameters have been determined considering multiple assumptions while the approximations for a few are not available. This makes rigorous comparison a challenging task. Nonetheless, dissimilarities in the tabulated features do represent certain key distinctions among different types of energy storage technologies.

In an electrical power network, the energy storage technology integrated to offer a distinct application can also be made available for many other potential benefits. The practicality and selection of a particular type of

Table 10.2 Role of ESS for multiple power system stakeholders.

Power system stakeholders	Role of energy storage system (ESS) technology
Generation and Distribution side	time shifting of energy output, seasonal energy storage (long term time shifting of energy), smoothing the fluctuating output of renewable energy technologies, maintains grid voltage stability, deferral of network expansion planning, mitigating reverse feedback, peak shaving, enhancing implementation of multiple demand response programs (DRPs), provide spinning reserve, support black start during power failure, substation and transformer capacity release, network voltage control support, energy arbitrage benefits, minimizing network investment and operation cost, enhancing network renewable hosting capacity, reducing carbon footprints, etc.
End-consumer side	satisfying energy and power demand, enhancing power reliability, less expensive backup power supply during short power outages, providing uninterrupted power supply, balancing supply and demand, providing capacity contract savings, proliferation of price and energy arbitrage benefits, enabling smart and sustainable homes, motivating and facilitating customer participation in different DRPs, etc.

energy storage technology are governed by the numerous factors and key technical characteristics which may include power rating, energy capacity, reaction time, frequency response, control and management strategy, lifespan, round-trip efficiency, charge/discharge power and duration, service requirement, investment cost, and many more. However, high initial investment cost is the key issue that restricts the implementation of most of these technologies in modern power networks [80]. Therefore, there is a pressing need for the development of these technologies with respect to performance enhancement and cost reduction except for pumped hydro energy storage.

Table 10.3 Important attributes of energy storage technology for specific applications in energy network [13, 14, 56].

Specific application	Facility size (MW)	Dispatch duration	Time to response
Voltage regulation	1-40	1 sec - 1 min	10^{-3} sec-sec
Frequency regulation	1-2000	1 min - 15 min	1 min
Renewable energy resource integration	1-400	1 min - 1 hour	less than 15 min
Spinning reserve	1-2000	15 min - 2 hours	less than 15 min
Time shifting of demand	0.001-0.01	1 min - hours	less than 15 min
Peak shaving	0.001-0.01	1 min - hours	less than 15 min
Backup supply and Black start	0.1-400	1 hour - 4 hours	less than 1 hour
Network congestion relief	10-500	2 hours - 4 hours	greater than 1 hour
Deferral of network expansion planning	1-500	2 hours - 5 hours	greater than 1 hour
Energy price arbitrage	100-2000	8 hours - 24 hours	greater than 1 hour
Seasonal storage	500-2000	days - months	a day

Several well-established and well-recognized storage technologies that offer an incredible economic solution to both the industrial and residential sector with significant storage size include pumped hydro energy storage and compressed air energy storage [81, 82]. They offer the highest power rating and energy capacity among all the energy storage technologies and are therefore mostly effective for large centralized storage. Due to the capability of storing energy for long periods, these storage technologies can effectively contribute to enhance the overall network energy efficiency and in time shifting of renewable power generation [78]. On the other hand, the optimal deployment of these technologies is restricted by certain factors like geographical concerns and complex infrastructural requirements. They also suffer from low round-trip efficiency [81, 83]. However, the authors in [83] concluded that for adiabatic compressed air energy storage the efficiency can be improved significantly.

Table 10.4 Key technical characteristics of the energy storage technologies [13, 14 57, 60 79].

Energy storage technology type	Efficiency = Sqrt (round-trip efficiency) (%)	Charge duration	Lifespan (years/number of cycles)	Power rating (MW)	Capacity (MWh)	Primary application	Capital cost Power ($/kW)	Capital cost Energy ($/kWh)
Compressed air energy storage (CAES)	0.50-0.89	1hour-months	20-60/8620-17100	5-300	14-2050	long-term storage, energy arbitrage	400-800	2-50
Pumped hydro energy storage (PHES)	0.60-0.85	1hour-months	40-60/12800-33000	100-5000	8000-190000	long-term storage	600-2000	5-100
Fuel cells (FCs)	0.2-0.66	1hour-months	5-15/1	0-50	fluctuates	long-term storage	>1000	--
Flywheel energy storage (FES)	0.85-0.95	1sec-5min	≤15/10^5-10^7	0-0.25	0.003-0.133	short-term storage	250-300	1000-5000
Super capacitor	0.84-0.97	1sec-hours	10-30/10^5-10^6	0-0.3	--	short-term storage	100-300	300-2000
Superconducting magnetic energy storage (SMES)	0.95-0.98	1min-4hours	>20/--	0.1-10	--	short-term storage	200-300	1000-10000

(Continued)

Table 10.4 Key technical characteristics of the energy storage technologies [13, 14 57, 60 79]. (Continued)

Energy storage technology type	Efficiency = Sqrt (round-trip efficiency) (%)	Charge duration	Lifespan (years/number of cycles)	Power rating (MW)	Capacity (MWh)	Primary application	Capital cost	
Battery Energy Storage System								
Lead acid	0.63-0.9	1min-days	5-15/160-1060	0-20	<30	distributed/off-grid storage, short-term storage	300-600	200-400
Vanadium redox (VR)	0.65-0.9	1hour-months	5-10/1510-2780	0.03-3	<10		600-1500	150-1000
Zinc bromide (ZnBr)	0.65-0.85	1hour-months	5-10/1510-2780	0.05-2	<10		700-2500	150-1000
Sodium sulphur (NaS)	0.75-0.9	1sec-5hours	10-15/1620-4500	0.05-8	7-450		1000-3000	300-500
Lithium ion (Li-ion)	0.75-0.97	1min-days	5-15/2960-5440	0-0.1	<22		1200-4000	600-2500
Metal air	0.5-0.55	1hour-months	--/--	0-0.01	--	--	100-250	10-60

Hydrogen energy storage or fuel cells represent another form of energy storage technology that can store surplus electricity from renewables. This type of energy storage provides numerous advantages in terms of high power and energy density, long charge durations and a highly compatible solution with existing conventional infrastructure for energy distribution in power networks. However, they suffer from lower efficiency and higher investment cost [84]. The higher investment cost may be significantly reduced by employing recent trends for hydrogen production using renewable power generation [85]. This may eventually provide an opportunity for economic energy storage. Despite being costly and having comparably lower efficiency, they are technically better and flexible for transport and other demanding services [78].

There are certain energy storage technologies that provide sufficiently high efficiency, rapid response time and fast frequency regulation ability. They may include superconducting magnetic energy storage (SMES), flywheel energy storage (FES), supercapacitor, and battery energy storage system (BESS). For charging and discharging of SMES and BESS, a power converter system (PCS) is used while it is normally employed in FES to provide a wider operating area to attached electrical machinery [86]. SMES services in electrical power network mainly depend upon the type of superconducting coil used for creating the magnetic field to store energy [87]. Even with high efficiency and ability to provide frequent response SMES requires high investment cost for its magnetic field generating part [86, 87]. This restricts the wide use of SMES in DNs. Due to growing integration of renewable energy technology in energy networks SMES is likely to offer a vital solution in assisting power quality and network stability enhancement [79]. FES may be used in conjunction with BESSs or sometimes may act as self-governing energy source when directly connected to DG. Regardless of lower maintenance cost, higher lifespan and immunity against the effects of depth of discharge (DOD) the restricted application feasibility of FES is the key concern due to high idling losses [87]. For low- and medium-term services in DN, FES may be utilized to offer rapid response frequency regulation [79] and therefore is beneficial to operate in combination with other storage technologies such as BESSs rather than offering individual backup supply. Supercapacitors are high energy density storage devices. The main advantages of supercapacitors include very low degradation cost, increased efficiency and higher life cycle. However, they are suggested only for short-term applications due to high daily self-discharge rate and investment cost [79, 87]. In modern power networks, supercapacitors may offer several services which may include energy smoothing and providing uninterrupted supply during voltage sags and brief interruption events, thus enhancing system reliability. Further, in certain specific applications high peak power and low energy are essential requirements. Under these circumstances, supercapacitors can act as potential candidates [86].

Now coming to the battery energy storage system (BESS), this represents a technically superior and versatile form of energy storage technology that can be employed in modern power networks. For independent short- and medium-term applications as well as for long-term applications, BESS deployment is recently inviting considerable attention from power system researchers [80]. Due to the contemporary technological advancements in BESS development, which results in decreased BESS price and enhanced capability to offer diverse potential benefits and advantages, the BESS utilization on a large scale is becoming the prime concern of multiple stakeholders in distribution network. However, to ensure its effective implementation and cost effective utilization, the operational-planning problems of BESSs must include technical and economic as well as environmental objectives. The numerous advantages offered by different types of BESSs include rapid response, large round-trip efficiency, longer life span, high power and energy density, easy deployment, fast charging and discharging ability, low maintenance requirement, wide operating temperature range, very low self-discharge, robust, and so on [88, 89]. The key components of any BESS technology include battery units, control mechanism and PCS [74, 90]. The several categories of BESSs that seem to be more appropriate for multiple power network applications may include lead acid, sodium sulphur (NaS), lithium ion (Li-ion), metal air and flow batteries. Flow batteries further comprise vanadium redox (VR) and zinc bromide (ZnBr) batteries.

The lead acid battery is believed to be the oldest and most well-established BESS technology among the aforesaid BESSs and has been employed in most power system applications [90]. Further, a lead acid battery can be flooded type or valve regulated (VRLA) type; the only difference is that in the case of VRLA a pressure regulating valve is used. For applications requiring bulk energy storage and rapid charging and discharging duration, the lead acid battery represents a highly feasible and economic choice to be pondered upon. However, the main disadvantages of this battery storage type include low energy density and shorter lifespan [86].

The NaS and Li-ion batteries are among the well-established battery technologies that show promising results in high power density application in DNs [90]. Among these two types, Li-ion is the most suitable for optimization problems and offers utmost potential for future advancement. Further, the Li-ion battery is highly suitable for convenient devices due to the compact size, reduced weight, highest efficiency (close to 100%) and energy density [80]. Despite being highly feasible and advantageous, the Li-ion battery suffers from high investment cost due to intricate fabrication of battery protection circuitry and reduced life span due to high depth of discharge. On the other hand, being smaller in size and lightweight, NaS suffers from high operating temperature range, personnel safety issues and also requires continuous heat input to be in a working state [87, 90]. The main issues with the metal air battery technology are the difficulty in recharging them once they are discharged and comparably lower efficiency; however, they offer potential viability in high energy density and low investment cost applications [13, 14].

For power network applications requiring long duration energy storages, flow batteries are technically perfect and promising BESS technologies due to their capability of not self-discharging, or we can say, they possess zero-self-discharge capability. The other advantages offered by flow batteries may include quick response and low maintenance [87, 90]. Contrarily, they are not suitable for short- and medium-term applications due to high investment and running cost. Further, the challenging part with the future advancement of flow battery technology is to enhance its power density significantly [13].

As already mentioned, in distribution networks BESS technologies can act as highly potential and viable candidates for multiple stakeholders to achieve profit in terms of enhanced energy price arbitrage benefits and other associated technical benefits due to wide range of applications like power network regulation and protection, balancing supply and demand, spinning and non-spinning reserve, power quality improvement, and many more. Among the said battery energy storage technologies, lead acid, Li-ion and NaS are the commonly used BESS technologies that have been extensively employed in distribution networks [13, 79].

10.5 Case Study

In this section, the authors established a multi-layer operation-planning joint optimization scheme for simultaneous joint allocation of multiple SPV and BES systems in contemporary distribution networks that ensure enhanced renewable hosting capacity, optimum utilization of deployed BESSs and increased system operational efficiency. In this case study, the proposed scheme encompasses a two-layered structure. The aim of the upper or outer layer is to determine the optimal siting-sizing of SPVs and BESSs simultaneously while minimizing multiple desired objectives. In the internal or inner layer, the optimum utilization of deployed BESSs is managed by suggesting some potential strategies for BESS operation while satisfying several network operation and BESS constraints. Further, a recently developed swarm-based algorithm, namely Modified African Buffalo Optimization (MABO) [10], having high exploration and exploitation capabilities is utilized to address the two-layer optimization problem. In this way, the suggested deterministic optimization structure essentially ensures significantly enhanced SPV hosting capacity of DN together with optimum utilization of deployed BESSs. The proposed optimization structure is implemented on a benchmark 33-bus test distribution network. The promising results of the case study highlight the importance of the proposed two-level optimization structure while effectively absorbing the fluctuating output of SPVs.

10.5.1 Proposed Two-Layer Optimization Framework and Problem Formulation

The renewable hosting capacity of vertically integrated distribution networks is limited due to the unidirectional network topology and passive nature. However, the rapid integration of multiple DER technologies has shifted the paradigm of passive distribution networks and resulted in the formation of active distribution networks (ADNs). Considering the above-discussed realities, the optimal allocation of multiple DERs may be crucial in neutralizing some of the high renewable integration impacts in DNs to a greater extent, although for DNOs the optimal allocation of multiple DERs is always proving to be a highly challenging task. The allocation problem emerges as a complex combinatorial, mixed-integer, non-linear and multi-constraint optimization problem while considering placement, sizing, number, and type of multiple DERs. The problem complexity further advances by integrating BESSs in DNs.

In this section, a two-layer operation-planning joint optimization model is developed so as to mitigate the problem complexity to some degree while optimally allocating BESSs and SPVs simultaneously. The proposed model is an intertwined structure composed of two layers, viz. upper and internal layer. In this model, the decision making at one layer is impacted by the decision making at another layer [91–94]. The working principle of this model lies in the fact that the internal layer optimization problem is treated as a constraint while solving the upper layer optimization problem. And the feasibility of optimal solution obtained at the internal layer is valid if and only if it satisfies the considered upper layer constraints [92–95]. The decision variables for the upper layer include placement and capacity of BESSs and SPVs while as for the internal layer hourly power dispatch of BESSs that further helps to estimate SOC acts as decision variables. Further, the model also involves interchange of variables from upper layer to internal layer and vice versa. These include optimal BESS and SPV location and capacity from upper layer to internal layer and optimal hourly BESS dispatch from internal layer to upper layer. Moreover, some potential management strategies for optimally utilizing deployed BESSs are developed in the internal layer of optimization model while taking care of network operation and BESS constraints, which are discussed at the end of this section. In each layer of the proposed optimization model, objective functions and constraints are formulated, which are presented as given below.

10.5.1.1 Upper-Layer Optimization

This layer provides the tentative planning solution of location and capacity of multiple SPVs and BESSs to the internal layer and therefore can also be termed as DER integration layer. The main contribution of this layer is to provide the final optimal capacity and placement of multiple SPVs and BESSs while minimizing several planning objectives. These objectives include minimization of annual energy losses (AEL) (f_1), load deviation index (LDI) (f_2), and daily charging and discharging energy mismatch index of BESSs (f_3). Since f_1 and f_2 are conflicting in nature and also all the three objectives are on a different scale. Therefore, to solve this multi-objective problem as a single objective optimization problem a penalty function–based method is employed [96]. However, the multi-objective problem can also be handled by employing a number of multi-objective optimization approaches which may include fuzzy approach [97, 98], weighted-sum approach [99, 100], and many more. The overall objective function for the upper layer can be mathematically formulated as given in Equation (10.1).

$$\min F_{upper} = (\Psi f_1)*(1+f_2)*(1+f_3) \qquad (10.1)$$

where,

$$f_1 = \sum_{i=1}^{N}\sum_{j=1}^{N} \alpha_{ij}^t (P_i^t P_j^t + Q_i^t Q_j^t) + \beta_{ij}^t (Q_i^t P_j^t - P_i^t Q_j^t); \alpha_{ij}^t = \frac{r_{ij}\cos(\delta_i^t - \delta_j^t)}{V_i^t V_j^t} \text{ and}$$

$$\beta_{ij}^t = \frac{r_{ij}\sin(\delta_i^t - \delta_j^t)}{V_i^t V_j^t}; \forall i,j \in \Omega_N, i \neq j, \forall t \in \Omega_T. \qquad (10.2)$$

$$f_2 = \left(\sqrt{\frac{1}{T}\sum_{t=1}^{T}\left(\overline{P}_G - P_G^t\right)^2}\right); \forall t \in \Omega_T \qquad (10.3)$$

$$f_3 = \max\left(\frac{\left|\sum_{t=1}^{T} P_{ch,i}^{BESS,t} - \sum_{t=1}^{T} P_{dis,i}^{BESS,t}\right|}{B_i^{rated}}\right); \forall i \in \Omega_N, \forall t \in \Omega_T \qquad (10.4)$$

In (10.1), Ψ denotes the daily to annual conversion factor. In (10.2), the notations used to denote real and reactive power injection, voltage magnitude and angle at ith node during tth system state and branch resistance between nodes i and j are P_i^t, Q_i^t, V_i^t, δ_i^t, and r_{ij}, respectively. In (10.3) and (10.4), $\overline{P}_G$, P_G^t, $P_{ch,i}^{BESS,t}$, $P_{dis,i}^{BESS,t}$, and B_i^{rated} denotes mean power generation from grid over time interval T, power generation from grid, charging and discharging dispatch of BESSs at ith node during tth system state and rated capacity of BESS to integrate at ith node. Further, in (10.2–10.4), i, j, N, Ω_N, t, T and Ω_T are representing system nodes, number and set of system nodes, system state, number of system states considered in a day and set of system states, respectively.

In this layer, the final optimal solution including location and capacity of SPVs and BESSs is determined. However, in each iteration the optional planning solution suggested by the outer layer needs to be justified. The objective functions defined by (10.2–10.4) are evaluated so as to find the fitness value of objective function defined by (10.1). On the other hand, for each system state $t \in T$ (where t is an integer), the time dependent variables involved in functions f_1 to f_3 need to be evaluated. Among these time dependent variables, daily charging and discharging dispatch of BESSs, i.e., $P_{ch,i}^{BESS}$ and $P_{dis,i}^{BESS}$ are optimization variables. Power flow analysis can be utilized to evaluate and update the remaining time dependent variables. However, at each system state t, the optimality of these variables is ensured by developing an internal layer optimization also called as operational layer optimization.

10.5.1.2 Internal-Layer Optimization

The purpose of the internal layer is to satisfy the operational constraints of the network and to ensure the optimal utilization of BESS and SPV capacities, suggested by the outer layer, at the time of planning itself. The objective functions considered in this layer include minimization of power delivery losses, and node voltage deviation (f_4). To solve these objective functions as a single objective function the same approach is adopted as suggested in the upper layer optimization framework. The overall objective function for this layer can be mathematically expressed as given in (10.5).

$$\min F_{internal} = (f_1)*(1+f_4) \tag{10.5}$$

where,

$$f_4 = \max \begin{pmatrix} 0; & if V^{min} \leq V_i^t \leq V^{max} \\ |1-V_i^t|; & else \end{pmatrix}; \forall i \in \Omega_N, \forall t \in \Omega_T \tag{10.6}$$

In (10.7), notations V^{min}, and V^{max} are denoting minimum and maximum specified node voltage limits.

10.5.1.3 Problem Constraints

The objective functions defined by (10.2–10.4), and (10.6) in both the upper and internal layer of optimization framework are subjected to the following constraints.

BESS operation limits

$$0 \leq B_i^{rated} \leq B^{max}; \forall i \in \Omega_N \tag{10.7}$$

$$P_{min}^{BESS} \leq P_{ch/dis,i}^{BESS,t} \leq P_{max}^{BESS}; \forall i \in \Omega_N, t \in \Omega_T \tag{10.8}$$

$$SOC^{min} \leq SOC_i^t \leq SOC^{max}; \forall i \in \Omega_N, t \in \Omega_T \tag{10.9}$$

$$SOC_i^t = SOC_i^{t-1} + \left(\frac{\sqrt{\eta} P_{ch,i}^{BESS,t}}{B_i^{rated}} - \frac{P_{dis,i}^{BESS,t}}{\sqrt{\eta} B_i^{rated}} \right) t; \forall i \in \Omega_N, t \in \Omega_T \tag{10.10}$$

$$P_{ch,i}^{UB,t} = \begin{cases} 0; & SOC_i^t = SOC^{max} \text{ or } \delta_1^t \geq \delta_2^t \\ P_{max}^{BESS}; & SOC_i^{t-1} + \frac{\sqrt{\eta} P_{max}^{BESS}}{B_i^{rated}} \Delta t < SOC^{max} \ \& \ \delta_1^t < \delta_2^t \quad ; \forall i \in \Omega_N, \forall t \in \Omega_T \\ (SOC^{max} - SOC_i^{t-1}) \frac{B_i^{rated}}{\Delta t}; & SOC^{max} - SOC_i^{t-1} < \frac{\sqrt{\eta} P_{max}^{BESS}}{B_i^{rated}} \Delta t \ \& \ \delta_1^t < \delta_2^t \end{cases} \tag{10.11}$$

$$P_{dis,i}^{UB,t} = \begin{cases} 0; & SOC_i^t = SOC^{min} \text{ or } \delta_1^t \le \delta_2^t \\ P_{min}^{BESS}; & SOC_i^{t-1} - \dfrac{P_{min}^{BESS}}{\sqrt{\eta} B_i^{rated}} \Delta t > SOC^{min} \, \& \, \delta_1^t > \delta_2^t \\ (SOC_i^{t-1} - SOC^{min}) \dfrac{W_B^R}{\Delta t}; & SOC_i^{t-1} - SOC^{min} < \dfrac{P_{min}^{BESS}}{\sqrt{\eta} B_i^{rated}} \Delta t \, \& \, \delta_1^t > \delta_2^t \end{cases} \quad (10.12)$$

Equation (10.7) denotes BESS capacity deployment limits at any node whereas (10.8) represents BESS charging and discharging power limits. Equations (10.9) and (10.10) are denoting State-of-Charge (SOC) limits and balancing, respectively. The maximum allowed BESS charging ($P_{ch,i}^{UB,t}$) and discharging ($P_{ch,i}^{UB,t}$) limits at ith node during state t are obtained by utilizing (10.11) and (10.12) respectively. The notations used to express maximum BESS capacity permitted to integrate at any node, minimum and maximum allowed charging and discharging power limits of BESS during any system state t, minimum and maximum SOC limits of BESS, SOC status of BESS at ith node during sth state, duration of state t, and round-trip efficiency of BESSs are B^{max}, P_{min}^{BESS}, P_{max}^{BESS}, SOC^{min}, SOC_i^t, SOC^{max}, Δt, and η respectively.

Power balance limits

$$P_i^t = V_i^t \sum_{j=1}^{N} V_j^t Y_{ij} \cos\left(\theta_{ij} + \delta_j^t - \delta_i^t\right); \forall i,j \in \Omega_N, i \ne j, \forall t \in \Omega_T \quad (10.13)$$

$$Q_i^t = -V_i^t \sum_{j=1}^{N} V_j^t Y_{ij} \sin\left(\theta_{ij} + \delta_j^t - \delta_i^t\right); \forall i,j \in \Omega_N, i \ne j, \forall t \in \Omega_T \quad (10.14)$$

In (10.13) and (10.14), Y_{ij} and θ_{ij} are denoting the Y-bus matrix features and impedance angle between ith and jth node.

SPV generation limits

$$0 \le P_{spv,i}^{rated} \le P_{spv}^{max}; \forall i \in \Omega_N \quad (10.15)$$

In (10.15), rated capacity of SPV installed at ith node and maximum capacity of SPV permitted to integrate at any node are denoted $P_{spv,i}^{rated}$ and

P_{spv}^{max} respectively. Further, it is to be noted that the modelling of SPV power generation is taken from [101].

Feeder thermal limits

$$I_{ij}^{t} \leq I_{ij}^{max}; \forall i, j \in \Omega_N, t \in \Omega_T \qquad (10.16)$$

In (10.16), current flow in branch during state t and maximum current limit of line between nodes i and j are denoted by I_{ij}^{t} and I_{ij}^{max} respectively.

Reverse feedback constraint

$$P_{RFB}^{t} = \begin{cases} \Re(V_1^t I_1^{t*}); & \delta_1^t < \delta_2^t \\ 0; & \delta_1^t \geq \delta_2^t \end{cases}, \forall t \in \Omega_T \qquad (10.17)$$

In (10.17), $\Re$, V_1^t, I_1^t, δ_1^t, δ_2^t are denoting network real power flow and grid supply point voltage, reverse feedback current, voltage angles at bus 1 and 2 during tth system state, respectively.

10.5.1.4 Proposed Management Strategies for BESS Deployment

The proposed strategies and assumptions implemented in this case study for optimal daily power and energy dispatch and management of deployed BESSs while framing the proposed two-layer operation-planning joint optimization model are listed below.

1. The charging and discharging of BESSs during excess renewable generation and on-peak hours, respectively, can act as a potential solution that may help to alleviate feeder power losses significantly.
2. The important factor that governs the sufficient renewable energy and BESS capacity integration in DNs is the reverse feedback, which is defined in (10.17).
3. Since BESSs are blessed with limited lifespan and number of dispatch cycles, only single dispatch cycle is considered for time interval T. Also, a minimum SOC level (SOC^{min}) of BESS is assumed to be 10% of rated BESS capacity and beyond this level no BESS is allowed to dispatch energy. These assumptions help to improve BESS lifespan and may also alleviate BESS conversion losses.
4. The SOC level at the beginning and end of a sample day should be equal; however, practically this may vary.

This constraint justifies the daily to annual operation as implemented in this chapter.

5. In order to mitigate the non-optimal utilization of deployed BESS suggested by the upper layer of the optimization model, it should be fully utilized in single dispatch cycle in a sample day.
6. The maximum upper and lower limits of hourly BESS charging and discharging power dispatch for any system state t for the internal layer of optimization model are evaluated by employing (10.11) and (10.12), respectively.

In the present case study, modified African buffalo optimization (MABO), having high exploration and exploitation capabilities, is employed to solve the proposed two-layer operation-planning joint optimization model while optimally allocating the multiple DERs, i.e., BESSs and SPVs. The ABO algorithm was first developed by Odili *et al.* in 2015 [102]. It is a swarm intelligence–based, nature-inspired, and meta-heuristic optimization algorithm motivated by the social and herding behaviour of African buffaloes [103]. It also offers the capability of solving optimization problems suffering from pre-mature convergence. On the other hand, it suffers from the inability to provide global optimum and to tackle complex engineering optimization problems. To overcome these limitations Singh *et al.* in January 2020 [10] developed a modified version of ABO named as MABO with the promising potential of offering global optimal solution for complex power system optimization problems. For further details like basic steps, proposed modifications in standard variant, mathematical equations and algorithm parameters of both ABO and MABO readers may refer to [102] and [10]. Moreover, the complete block diagram of the proposed optimization model, utilizing MABO in both the upper and internal layer, for operational planning of multiple BESSs and SPVs is shown in Figure 10.1.

10.5.2 Results and Discussions

In this section, the developed joint optimization model is authenticated by realizing on a benchmark test DN consisting of 33 buses [104]. The considered test system is a 12.66 kV primary DN with the loading of 3715 kW active demand and 2300 kVAr reactive demand. At nominal loading, the minimum node voltage and real power losses of this system are 0.9131 p.u. and 202.67 kW, respectively. The bus and line data of this system is presented in Table A.10.1 of Appendix A. For additional details regarding this test system, the reader may refer to [104]. Several practical parameters

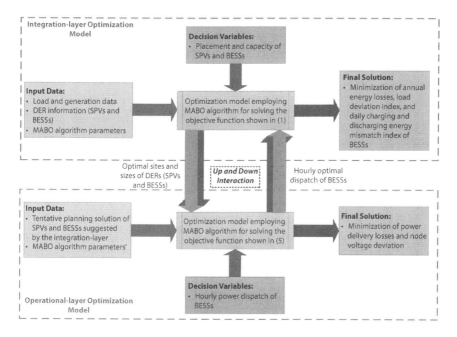

Figure 10. 1 Block diagram of theproposedoptimization model.

considered while simulating the proposed joint optimization model are listed in Table 10.5. In this case study, the hourly multiplying factors (MFs) for solar power and load demand for a sample day are taken from [105], as shown in Figure 10.2. Moreover, the overall network demand is assumed to be composed of residential, commercial and industrial end-consumers as proposed in [106]. The diversity in loading pattern of the different types of end-consumers provides a closely related practical scenario of DNs. The hourly load demand and solar power generation in kW and kWp are evaluated by multiplying the respective MFs with the nominal system demand and SPVs rated capacity, respectively. Further, in accordance with industrial know-how, the SPVs are only accessible in certain specific dimensions and consequently the present case study only considers discrete capacities of SPVs. The proposed model is solved by employing MABO algorithm with population size of 50 and maximum generation count of 100 for both the integration layer and operation layer. For the rest of the parameters concerning MABO algorithm, Ref. [10] may be referred. The application results of the proposed method for optimal location and capacity of SPVs and BESSs are shown in Table 10.6. The load flow analysis is performed by utilizing the Backward/Forward sweep load flow method. Further, Li-ion

Table 10.5 Practical parameters employed in the case study.

Practical parameter(s)	Numerical value(s)
Ψ	365 days
T	24 states
η	85%
V^{min}, V^{max}	0.95 p.u., 1.05 p.u.
$P_{min}^{BESS}, P_{max}^{BESS}$	1000 kW each
SOC^{min}, SOC^{max}	0.1, 1
P_{spv}^{max}, B^{max}	2000 kWp, 5000 kWh

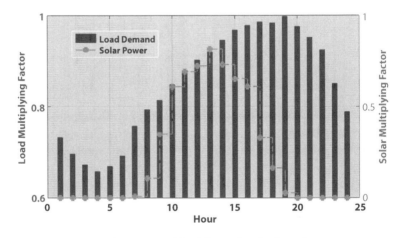

Figure 10.2 State-wise solar power and load demand multiplying factors.

battery technology is adopted in this case study due to high feasibility for optimization problems and highest efficiency and energy density [80].

In order to examine the efficacy and applicability of the proposed joint optimization model three cases are explored:

Case-I: *Base case (without any DER)*
Case-II: *Only SPVs integration*
Case-III: *Proposed joint integration of SPVs and BESSs simultaneously*

Table 10. 6 Optimal location and capacity of SPVs and BESSs.

DER nodes	DER capacities	
	SPV (kWp)	BESS (kWh)
17	1700	4100
24	-	610
25	625	-
33	900	1940

The application results obtained after simulating the above-mentioned cases are summarized in Table 10.7. The table contains optimal locations and capacities of SPVs and BESSs, level of SPV penetration (evaluated as a fraction of system's peak demand which in the present case study is 3715 kW), annual energy loss (AEL) in the system, load demand deviation, minimum and mean voltage, reduction in AEL, and enhancement in LDI. The load demand profile and solar power generation utilized for simulation purpose of cases I, II and III are shown in Figure 10. 2. From the analysis of the data given in Table 10.7, it can be observed that the case-III outperforms the remaining two cases i.e., case-I and case-II.

In case-I, also called as base case, not a single DER is considered while performing the power flow calculations using the Backward/Forward sweep load flow method on load profile as shown in Figure 10. 2. In this case, the minimum of mean voltage, annual feeder energy losses, daily feeder power losses and load deviation are found to be 0.913 p.u., 1279.60 MWh, 3505.80 kW, and 467.30 kW, respectively. Further, in this case the mean node voltage profile of the DN is significantly below the predefined limit.

In case-II, the optimal allocation of only SPV units is determined while optimizing the proposed joint optimization model. According to Ref. [107], only three SPV units are found to be near optimal for 33-bus test DN. Therefore, in this case only three SPVs are supposed to be placed in the DN. The internal layer optimization model is not evaluated here because only non-dispatchable sources (SPVs) are integrated in this case. Although there is a significant reduction in AEL by about 17.91%, possibly due to the placement of two SPV units on nodes 18 and 30, there is no significant improvement in load deviationas compared to the base case, as can be observed from Table 10.7, which is a big concern for the network operators. This can possibly be due to the fluctuating nature of power generation from SPVs which needs a flexible source for absorbing the fluctuations and

Table 10.7 Simulation results of the proposed joint optimization model attained by employing modified African buffalo optimization (MABO) algorithm.

Case (s)	Bus location for SPV placement	SPV capacity (kWp)	SPV penetration (%)	Bus location for BESS placement	BESS capacity (kWh)	AEL (MWh)	Load deviation (kW)	Min. of Mean Voltage (p.u.)	Reduction in AEL (%)	Improvement in LDI (%)
I	-	-	0.00	-	-	1279.60	467.30	0.913	0.00	0.00
II	18 30 21	570 465 300	36.00	-	-	1050.35	446.41	0.94	17.91	4.47
III	17 25 33	1700 625 900	86.81	17 24 33	4100 610 1940	930.18	203.16	0.96	27.30	56.52

delivering the same when the system is working under stringent conditions. For clear understanding the reader may refer to Figure 10.3. Further, the maximum power generation from SPV units does not coincide with the peak load demand, as is evident from Figure 10.2. This restricts the penetration limit of SPVs to only 36% in this case and also confines the renewable hosting capacity of the network. Moreover, the mean node voltage profile is improved in comparison to the case-I, but the minimum value is still under the predefined lower limit, as can be observed from Figure 10.4. Considering the above-mentioned issues and to enhance the overall operational efficiency and performance of ADNs the allocation of SPVs is re-investigated in case-III by optimally allocating three SPVs and three BESSs simultaneously in the present case study. It can be observed from

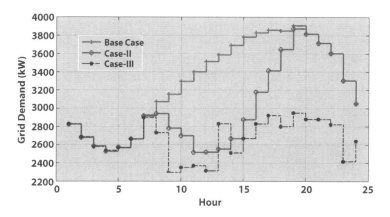

Figure 10.3 Comparison of hourly grid demand profiles of the system.

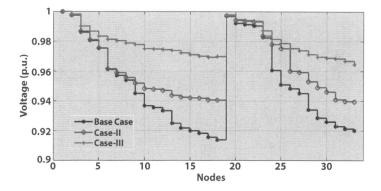

Figure 10.4 Comparison of mean voltage profiles of the system.

the data given in Table 10.7 that the case-III unveils the possible potential techno-economic benefits of BESSs in DNs when deploying simultaneously in combination with SPVs.

The integration of BESSs in case-III has increased the capacity of all the three SPV units, thereby increasing the SPV penetration to 86.81%. In other words, we can say that the renewable hosting capacity of DN is significantly enhanced. With this SPV penetration, annual energy losses are reduced from 1279.60 to 930.18 MWh, i.e., by about 27%, and load deviation index is improved by 56.52%, i.e., from 467.30 kW to 203.16 kW, as can be observed from Table 10.7. To gain these benefits and to facilitate this sufficiently high renewable penetration 6650 kWh of energy storage technology is deployed optimally. From this table, it can be observed that the mean node voltage profile is considerably enhanced as compared to case I and II, having a minimum value of 0.96 p.u, which is also evident from Figure 10.4. It is important to notice that the optimal placement of two SPVs and two BESSs in this case occurs on the same nodes while the third SPV and BESS are placed on consecutive nodes. This possibly happens so as to mitigate the system losses while transacting energy among these DERs and also with system.

In case-III, the obtained optimal solution for the hourly power dispatch and state-of-charge (SOC) status of multiple BESS units placed on nodes 17, 24 and 33 for 24 states are presented in Figure 10.5 and Figure 10.6, respectively. The charging of BESSs mostly coincides with the SPV power generation profile from 9:00 to 17:00 Hours with the exception of one BESS at node 33 which discharges at 17:00 Hours, as is clear from Figure 10.5 and Figure 10.2. Therefore, most of the fluctuating power generation

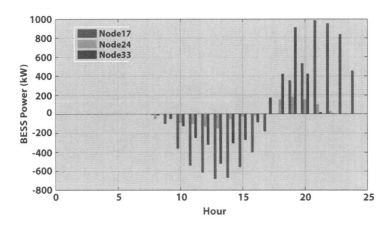

Figure 10.5 Optimal solution for the hourly power dispatch of multiple BESS units.

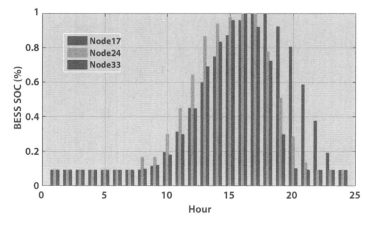

Figure 10.6 Optimal solution for the hourly SOC status of multiple BESS units.

from SPVs is absorbed and made dispatchable by the deployed BESSs. This also facilitates the mitigation of reverse feedback or power flow when the power generation from renewables occurs in excess. Also, the BESSs inject energy into the system mostly during highly stressed working conditions or peak load hours which in turn helps in clipping the peak demand on the grid, which is obvious from Figure 10.3, Figure 10.5 and Figure 10.6. This tapping of renewable power generation and then delivering to the system when most needed fulfills the BESS deployment strategies already discussed in Section 5. Further, the SOC status of deployed BESSs, as shown in Figure 10.6, falls within the prescribed limits at all the optimal locations. Therefore, it can be concluded that the deployed BESS units have been optimally utilized under sufficiently high SPV penetration while satisfying system operation and BESS constraints and suggested BESS deployment strategies faithfully.

The proposed simultaneous joint optimization of SPVs and BESSs significantly impacts the grid demand. Despite high SPV penetration, the load deviation index is highly enhanced as shown in Figure 10.3. Figure 10.3 presents the comparison of hourly grid demand profiles for cases I, II and III. Similarly, the comparison of mean voltage profiles of the system for all the three cases is shown in Figure 10.4. From Figure 10.3 and Figure 10.4, it can be observed that the system load demand and mean voltage profiles are significantly improved in case-III, respectively, as compared to cases I and II. This certainly relieves the system from harsh working circumstances and may also enhance the network reliability, efficacy, power quality and self-adequacy. Also the net grid demand is well above the horizontal axis showing the total mitigation of reverse feedback. Moreover, the percentage

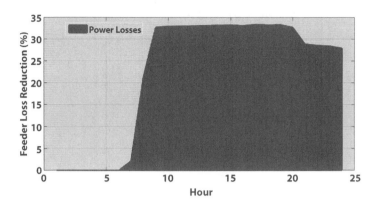

Figure 10.7 Percentage power loss reduction of the system.

power loss reduction of the system while optimally allocating SPVs and BESSs simultaneously is shown in Figure 10.7. The figure indicates that throughout the operation period of these DERs, the hourly feeder power losses are significantly reduced by about 29%, which amounts to around 349.42 MWh annually. These promising results highlight the importance of the proposed optimization model in enhancing distribution network performance and renewable hosting capacity while optimizing all the proposed objectives positively.

10.5.3 Conclusions

In this case study, a two-layer operation-planning joint optimization model is developed for simultaneously allocating and managing multiple SPV and BESS units in active distribution networks while ensuring enhanced system performance and renewable generation hosting capacity and deployment of suggested BESSs optimally. Further, the proposed BESS deployment strategies effectively absorb the fluctuating renewable power generation and also reduce the grid demand fluctuations. The strategic allocation of BESSs in combination with SPVs reveals the possible potential benefits of BESS units in modern power systems while optimizing several network operator objectives successfully. Finally, the application results attained confirms that the proposed optimization model efficiently and competently manages to allocate the maximum solar power generation in distribution network while satisfying numerous system operation and BESS constraints without any violations.

10.6 Future Research and Recommendations

In a broader outlook, energy storage technologies can act as feasible facilities for network operator, which is obvious from the reviewed literature, with highly intermittent and variable supply and demand side. Numerous literary works, which are unquestionable, have been presented determining the optimal allocation and operation of these technologies along with renewable technologies in solving multiple complex and combinatorial network operator problems. However, the rapid changing of power system requirements with the transition of existing networks to low carbon networks requires a deeper analysis to explore the possible potential benefits of these technologies. Further, to improve the capability and operation performance of storage technologies there is a pressing need to develop multiple operation and management strategies and to understand how and where these strategies and particular storage technology can bring surplus benefits to the DNOs.

The rapid integration of renewable technologies in distribution networks results in a series of issues and challenges that may further restrict the renewable generation hosting capacity of the DNs. The energy storage technology can act as a potential candidate in renewable integrated DNs. Therefore, there is a need to develop a framework for simultaneously allocating energy storage and renewable technologies to alleviate some of these issues and challenges and to enhance the renewable hosting capacity of distribution networks. Following this, the authors have developed a multilayer operation-planning joint optimization model already discussed and solved in the case study section of this chapter. However, the proposed joint optimization model may be extended in future to investigate the economic aspects of SPVs and BESSs in DNs and impact analysis of EVs and DR programs on distribution network performance.

Since energy storage technology represents a costly affair despite providing profit gains through energy and price arbitrage, integration of storage technology in modern power system needs to be justified economically. The cost-effective integration of storage technologies may be justified to a greater extent if these ensure sufficiently high penetration of renewable technologies along with other specified potential benefits. Moreover, there is a growing need for the cost-effective development of these technologies.

The various parameters that affect the energy storage technology economics may include selection of storage technology and its application, power and energy rating, power system structure, and so on. These factors make the estimation of storage technology economics a very complex problem. Also, the planning and operation of these technologies is

Appendix A

Table A.10.1 Line and Bus data of 33-bus test distribution system.

Br. no.	Sen. end	Rec. end	R (Ω)	X (Ω)	Rec. end load kW	Rec. end load kVAr	Br. no.	Sen. end	Rec. end	R (Ω)	X (Ω)	Rec. end load kW	Rec. end load kVAr
1	1	2	0.0922	0.047	100	60	17	17	18	0.732	0.574	90	40
2	2	3	0.493	0.2511	90	40	18	2	19	0.164	0.1565	90	40
3	3	4	0.366	0.1864	120	80	19	19	20	1.5042	1.3554	90	40
4	4	5	0.3811	0.1941	60	30	20	20	21	0.4095	0.4784	90	40
5	5	6	0.819	0.707	60	20	21	21	222	0.7089	0.9373	90	40
6	6	7	0.1872	0.6188	200	100	22	3	23	0.4512	0.3083	90	50
7	7	8	0.7114	0.2351	200	100	23	23	24	0.898	0.7091	420	200
8	8	9	1.03	0.74	60	20	24	24	25	0.898	0.7011	420	200
9	9	10	1.044	0.74	60	20	25	6	26	0.203	0.1034	60	25
10	10	11	0.1966	0.065	45	30	26	26	27	0.2842	0.1447	60	25
11	11	12	0.3744	0.1238	60	35	27	27	28	1.059	0.9337	60	20
12	12	13	1.468	1.155	60	35	28	28	29	0.8042	0.7006	120	70
13	13	14	0.5416	0.7129	120	80	29	29	30	0.5075	0.2585	200	600
14	14	15	0.591	0.526	60	10	30	30	31	0.9744	0.963	150	70
15	15	16	0.7463	0.545	60	20	31	31	32	0.3105	0.3619	210	100
16	16	17	1.289	1.721	60	20	32	32	33	0.341	0.5302	60	40

embedded and cannot be performed separately [14]. This further adds to the complexity of the problem and makes the optimization process computationally highly intensive. Therefore, the development of a well-tailored and well-coordinated multi-level optimization model or framework with the inclusion of recently explored swarm intelligence–based or evolutionary algorithms is the need of the hour.

Acknowledgement

This publication is supported by Visvesvaraya PhD Scheme, MeitY, Govt. of India VISPHD-MEITY-1981.

References

1. Gao, S., Liu, S., Liu, Y., Zhao, X., Song,T. E., Flexible andeconomic dispatching of ac/dc distribution networks consideringuncertainty of wind power. *IEEE Access*, 7, 100051-100065, 2019.
2. Xiao, J., *et al.*, Flexible distribution network: definition, configuration, operation, and pilot project. *IET Generation, Transmission & Distribution*, 12, 20, 4492-4498, 2018.
3. Bruce, A. R. W., Gibbins, J., Harrison G. P., ChalmersH.,Operational flexibility of future generation portfolios using highspatial- and temporal-resolution wind data. *IEEE Transactions Ion Sustainable Energy*, 7, 2, 697-707, 2016.
4. Jannesar, M. R., Sedighi, A., Savaghebi, M., Guerrero, J. M., Optimal placement, sizing, and daily charge/discharge of battery energy storage in low voltage distribution network with high photovoltaic penetration. *Applied Energy*, 226, 957-966, 2018.
5. Wong, L. A., Ramachandaramurthy, V. K., Walker, S. L., Taylor, P., Sanjari, M. J., Optimal placement and sizing of battery energy storage system for losses reduction using whale optimization algorithm. *Journal of Energy Storage*, 26, 100892, 2019.
6. Luo, L., Abdulkareem, S. S., Rezvani, A., Miveh, M. R., Samad, S., Aljojo, N., Pazhoohesh, M., Optimal scheduling of a renewable based microgrid considering photovoltaic system and battery energy storage under uncertainty. *Journal of Energy Storage*, 28, 101306, 2020.
7. Xiao, J., Zhang, Z., Bai, L., Liang, H., Determination of the optimal installation site and capacity of battery energy storage system in distribution network integrated with distributed generation. *IET Generation, Transmission & Distribution*, 10, 3, 601-607, 2016.

8. Ahmadi, M., Adewuyi, O. B., Danish, M. S. H., Mandal, P., Yona, A., Senjyu, T., Optimum coordination of centralized and distributed renewable power generation incorporating battery storage system into the electric distribution network. *International Journal of Electric Power and Energy Systems*, 125, 106458, 2021.
9. Kumar, A., Meena, N. K., Singh, A. R., Deng, Y., He, X., Bansal, R., Kumar, P., Strategic integration of battery energy storage systems with the provision of distributed ancillary services in active distribution systems. *Applied Energy*, 253, 113503, 2019.
10. Singh, P., Meena, N. K., Slowik, A., Bishnoi, S. K., ModifiedAfrican buffalo optimization for strategic integration of battery energy storage in distribution networks. *IEEE Access*, 8, 14289-14301, 2020.
11. Thokar, R. A., Gupta, N., Niazi, K. R., Swarnkar, A., Meena,N. K., Estimation of State-of-Charge for Dynamic Management of Battery Energy Storage Systems for High Renewable Penetration. *2020 IEEE International Conference on Power Electronics & IOT Applications in Renewable Energy and its Control (PARC)*, Mathura, Uttar Pradesh, India, 96-101, 2020.
12. Yang, Y., Li, H., Aichhorn, A., Zheng, J., Greenleaf, M., Sizing strategy of distributed battery storage system with high penetration of photovoltaic for voltage regulation and peak load shaving. *IEEE Transactions on Smart Grid*, 5, 982-991, 2014.
13. Wong, L. A., Ramachandaramurthy, V. K., Taylor, P., Ekanayake, J., Walker, S. L., Padmanaban,S., Review on the optimal placement, sizing and control of an energy storage system in the distribution network. *Journal of Energy Storage*, 21, 489–504, 2019.
14. Zidar, M., Georgilakis, P. S., Hatziargyriou, N. D., Capuder, T., Škrlec, D., Review of energy storage allocation in power distribution networks: applications, methods and future research. *IET Generation, Transmission & Distribution*, 10, 3, 645-652, 2016.
15. Sheibani, M. R., Yousefi, G. R., Latify, M. A., Dolatabadi, S. H., Energy storage system expansion planning in power systems: a review. *IET Generation, Transmission & Distribution*, 12, 11, 1203-1221, 2018.
16. Das, C. K., Bass, O., Kothapalli, G., Mahmoud, T. S., Overview of energy storage systems in distribution networks: Placement, sizing, operation, and power quality. *Renewable and Sustainable Energy Reviews*, 91, 1205-1230, 2018.
17. Singh, P., Meena, N. K., Yang, J., Vega-Fuentes, E., Bishnoi, S. K., Multi-criteria decision making monarch butterfly optimization for optimal distributed energy resources mix in distribution networks. *Applied Energy*, 278, 115723, 2020.
18. Sedghi, M., Ahmadian, A., Aliakbar-Golkar, M., Optimal storage planning in active distribution network considering uncertainty of wind power distributed generation. *IEEE Transactions on Power Systems*, 31, 1, 304–316, 2016.

19. Zhang, Y., Meng, K., Luo, F., Dong, Z. Y., Wong, K. P., Zheng, Y., Optimal allocation of battery energy storage systems in distribution networks with high wind power penetration. *IET Renewable Power Generation*, 10, 8, 1105–1113, 2016.
20. Karanki, S. B., Xu, D., Optimal capacity and placement of battery energy storage systems for integrating renewable energy sources in distribution system. *Proc. Nat. Power Syst. Conf. (NPSC)*, 1–6, 2016.
21. Li, Y., Feng, B., Li, G., Qi, J., Zhao, D., Mu, Y., Optimal distributed generation planning in active distribution networks considering integration of energy storage. *Applied Energy*, 210, 1073–1081, 2018.
22. Oldewurtel, F., *et al.*, A framework for and assessment of demand response and energy storage in power systems. *2013 IREP Symposium Bulk Power System Dynamics and Control - IX Optimization, Security and Control of the Emerging Power Grid*, Rethymno, 1-24, 2013.
23. Lo, C., Anderson, M., Economic dispatch and optimal sizing of battery energy storage systems in utility load-leveling operations. *IEEE Transactions on Energy Conversion*, 14, 3, 824–829, 1999.
24. Li, Q., Choi, S. S., Yuan, Y., Yao, D. L., On the determination of battery energy storage capacity and short-term power dispatch of a wind farm. *IEEE Transactions on Sustainable Energy*, 2, 2, 148-158, 2011.
25. Wilson, I. G., McGregor, P. G., Infield, D. G., Hall, P. J., Grid-connected renewables, storage and the UK electricity market. *Renewable Energy*, 36, 8, 2166–2170, 2011.
26. Swift-Hook, D. T., Wind energy really is the last to be stored and solar energy cannot be stored economically. *Renewable Energy*, 50, 971–976, 2013.
27. Wang, X., Determination of battery storage capacity in energy buffer for wind farm. *IEEE Transactions on Energy Conversion*, 23, 3, 868–878, 2008.
28. Yao, D. L., Choi, S. S., Tseng,K. J., Lie, T. T., A statistical approach to the design of a dispatchable wind power-battery energy storage system. *IEEE Transactions on Energy Conversion*, 24, 4, 916-925, 2009.
29. Kaldellis, J. K., Zafirakis, D., Kondili, E., Optimum sizing of photovoltaic-energy storage systems for autonomous small islands. *International Journal of Electric Power and Energy Systems*, 32, 1, 24-36, 2010.
30. Mitra, J., Reliability-based sizing of backup storage. *IEEE Transactions on Power Systems*, 25, 2, 1198-1199, 2010.
31. Bludszuweit, H., Domínguez-navarro, J. A., A probabilistic method for energy storage sizing based on wind power forecast uncertainty. *IEEE Transactions on Power Systems*, 26, 3, 1651–1658, 2011.
32. Makarov, Y., Du, P., Sizing energy storage to accommodate high penetration of variable energy resources. *IEEE Transactions on Sustainable Energy*, 3, 1, 34-40, 2012.

33. Mitra, J., Vallem, M., Determination of storage required to meet reliabilityguarantees on Island-capable microgrids with intermittent sources. *IEEE Transactions on Power Systems*, 27, 4, 2360-2367, 2012.
34. Hartmann, B., Dán, A., Methodologies for storage size determination for the integration of wind power. *IEEE Transactions on Sustainable Energy*, 5, 1, 182–189, 2014.
35. Wang, S. Y., Yu,J. L., Optimal sizing of the CAES system in a power system with high wind power penetration. *International Journal of Electric Power and Energy Systems*, 37, 1, 117–125, 2012.
36. Venu, C., Riffonneau, Y., Bacha, S., Baghzouz, Y., Battery storage system sizing in distribution feeders with distributed photovoltaic systems. *2009 IEEE Bucharest PowerTech.*, Bucharest, 1-5, 2009.
37. Abbey, C., Joòs, G., A stochastic optimization approach to rating of energy storage systems in wind-diesel isolated grids. *IEEE Transactions on Power Systems*, 24, 1, 418-426, 2008.
38. Atwa, Y. M., El-Saadany, E., Optimal allocation of ess in distribution systems with a high penetration of wind energy. *IEEE Transactions on Power Systems*, 25, 4, 1815-1822, 2010.
39. Pudjianto, D., Aunedi, M., Djapic, P., Strbac, G., Whole-systems assessment of the value of energy storage in low-carbon electricity systems. *IEEE Transactions on Smart Grid*, 5, 2, 1098 1109, 2014.
40. Nick, M., Cherkaoui, R., Paolone, M., Optimal allocation of dispersed energy storage systems in active distribution networks for energy balance and grid support. *IEEE Transactions on Power Systems*, 29, 5, 2300–2310, 2014.
41. Zidar, M., Capuder, T., Georgilakis, P. S., Škrlec, D., Convex AC optimal power flow method for definition of size and location of battery storage systems in the distribution grid. *Proc. of the Ninth Conf. on Sustainable Development of Energy, Water and Environment System (SDEWES)*, 1–23, 2014.
42. Tsung-Ying, L., Nanming C., Determination of optimal contract capacities and optimal sizes of battery energy storage systems for time-of-use rates industrial customers. *IEEE Transactions on Energy Conversion*, 10, 3, 562-568, 1995.
43. Oh, H., Optimal planning to include storage devices in power systems. *IEEE Transactions on Power Systems*, 26, 3, 1118-1128, 2011.
44. Akhavan-Hejazi, H., Mohsenian-Rad, H., Optimal operation of independent storage systems in energy and reserve markets with high wind penetration. *IEEE Transactions on Smart Grid*, 5, 2, 1088-1097, 2014.
45. Borowy, B. S., Salameh, Z. M., Methodology for optimally sizing the combination of a battery bank and PV array in a wind/PV hybrid system. *IEEE Transactions on Energy Conversion*, 11, 2, 367-375, 1996.
46. Mohammadi, S., Mozafari, B., Solymani, S., Niknam, T., Stochastic scenario-based model and investigating size of energy storages for PEM-fuel

cell unit commitment of micro grid considering profitable strategies. *IET Generation, Transmission & Distribution*, 8, 7, 1228-1243, 2014.
47. Chen, C., Duan, S., Cai, T., Liu,B., Hu, G., Optimal allocation and economic analysis of energy storage system in microgrids. *IEEE Transactions on Power Electronics*, 26, 10, 2762-2773, 2011.
48. Carpinelli, G., Mottola, F., Proto, D., Russo, A., Optimal allocation of dispersed generators, capacitors and distributed energy storage systems in distribution networks. *2010 Modern Electric Power Systems*, Wroclaw, 1-6, 2010.
49. Ghofrani, M., Arabali, A., Etezadi-Amoli, M., Fadali, M. S., A framework for optimal placement of energy storage units within a power system with high wind penetration. *IEEE Transactions on Sustainable Energy*, 4, 2, 434-442, 2013.
50. Sedghi, M., Aliakbar-Golkar, M., Haghifam, M., Distribution network expansion considering distributed generation and storage units using modified PSO algorithm. *International Journal of Electric Power and Energy Systems*, 52, 221-230, 2013.
51. Kahrobaee, S., Asgarpoor, S., Qiao, W., Optimum sizing of distributed generation and storage capacity in smart households. *IEEE Transactions on Smart Grid*, 4, 4, 1791-1801, 2013.
52. Zheng, Y., Dong, Z. Y., Luo, F. J., Meng, K., Qiu, J., Wong, K. P., Optimal allocation of energy storage system for risk mitigation of discos with high renewable penetrations. *IEEE Transactions on Power Systems*, 29, 1, 212-220, 2014.
53. Mukhopadhyay, B., Das, D., Multi-objective dynamic and static reconfiguration with optimized allocation of PV-DG and battery energy storage system. *Renewable and Sustainable Energy Reviews*, 124, 109777, 2020.
54. Jamian, J. J., Mustafa, M. W., Mokhlis, H., Baharudin, M., Simulation study on optimal placement and sizing of battery switching station units using Artificial Bee Colony algorithm. *International Journal of Electric Power and Energy Systems*, 55, 592-601, 2014.
55. Bahmani-Firouzi, B., Azizipanah-Abarghooee, R., Optimal sizing of battery energy storage for micro-grid operation management using a new improved bat algorithm. *International Journal of Electric Power and Energy Systems*, 56, 42–54, 2014.
56. Lott, M., Kim, S., Tam, C., Elzinga, D., Heinen, S., Munuera, L., Remme, U., (2014). Technology Roadmap: Energy storage.
57. Díaz-González, F., Sumper, A., Gomis-Bellmunt, O., Villafáfila-Robles, R., A review of energy storage technologies for wind power applications. *Renewable and Sustainable Energy Reviews*, 16, 4, 2154-2171, 2012.
58. Zhao, H., Wu, Q., Hu, S., Xu, H., Rasmussen, C.N., Review of energy storage system for wind power integration support. *Applied Energy*, 137, 545-553, 2015.

59. Gallo, A., Simões-Moreira, J., Costa, H., Santos, M., dos Santos, E. M., Energy storage in the energy transition context: a technology review. *Renewable and Sustainable Energy Reviews*, 65, 800-822, 2016.
60. Chen, H., Cong, T.N., Yang, W., Tan, C., Li, Y., Ding, Y., Progress in electrical energy storage system: a critical review. *Progress in Natural Science*, 19, 3, 291-312, 2009.
61. Arul, P., Ramachandaramurthy, V. K., Mitigating techniques for the operational challenges of a standalone hybrid system integrating renewable energy sources. *Sustainable Energy Technologies and Assessments*, 22, 18–24, 2017.
62. Denholm, P., Hand, M., Grid flexibility and storage required to achieve very high penetration of variable renewable electricity. *Energy Policy*, 39, 3, 1817-1830, 2011.
63. Wade, N.S., Taylor, P.C., Lang, P.D., Jones, P.R., Evaluating the benefits of an electrical energy storage system in a future smart grid. *Energy Policy*, 38, 11, 7180-7188, 2010.
64. Ekren, O., Ekren, B. Y., Ozerdem, B., Break-even analysis and size optimization of a PV/wind hybrid energy conversion system with battery storage – a case study. *Applied Energy*, 86, 7-8, 1043-1054, 2009.
65. Ekren, O., Ekren, B. Y., Size optimization of a PV/wind hybrid energy conversion system with battery storage using simulated annealing. *Applied Energy*, 87, 2, 592-598, 2010.
66. Khani, H., Zadeh, M. R. D., Hajimiragha, A. H., Transmission congestion relief using privately owned large-scale energy storage systems in a competitive electricity market. *IEEE Transactions on Power Systems*, 31, 2, 1449-1458, 2016.
67. Brekken, T. K. A., Yokochi, A., Von Jouanne, A.Z., Yen, Z., Hapke, H. M., Halamay, D. A., Optimal energy storage sizing and control for wind power applications. *IEEE Transactions on Sustainable Energy*, 2, 1, 69-77, 2011.
68. Le, H. T., Nguyen, T. Q., Sizing energy storage systems for wind power firming: An analytical approach and a cost-benefit analysis. *2008 IEEE Power and Energy Society General Meeting - Conversion and Delivery of Electrical Energy in the 21st Century*, Pittsburgh, PA, 1-8, 2008.
69. Fallahi, F., Nick, M., Riahy, G. H., Hosseinian, S. H., Doroudi, A., The value of energy storage in optimal non-firm wind capacity connection to power systems. *Renewable Energy*, 64, 34-42, 2014.
70. Gyuk, I. P., Eckroad, S., *Energy storage for grid connected wind generation applications*, (US Department of Energy, Washington, DC, 2004), EPRI-DOE Handbook Supplement, 1008703, 2004.
71. Chen, S., Gooi, H., Wang, M., Sizing of energy storage for microgrids. *IEEE Transactions on Smart Grid*, 3, 1, 142-151, 2012.
72. Hatta, H., Asari, M., Kobayashi, H., Study of energy management for decreasing reverse power flow from photovoltaic power systems. *2009 IEEE PES/IAS Conference on Sustainable Alternative Energy (SAE)*, Valencia, 1-5, 2009.

73. Wen, Y., Guo, C., Pandžić H., Kirschen, D. S., Enhanced security-constrained unit commitment with emerging utility-scale energy storage. *IEEE Transactions on Power Systems*, 31, 1, 652-662, 2016.
74. Suberu, M. Y., Mustafa, M. W., Bashir, N., Energy storage systems for renewable energy power sector integration and mitigation of intermittency. *Renewable and Sustainable Energy Reviews*, 35, 499-514, 2014.
75. Thokar, R. A., Gupta, N., Niazi, K. R., Swarnkar, A., Meena, N. K., A Coordinated Operation of Multiple Distributed Energy Resource Technologies for Arbitrage Benefit Enhancement. *2020 IEEE International Conference on Power Electronics & IOT Applications in Renewable Energy and its Control (PARC)*, Mathura, Uttar Pradesh, India, 143-147, 2020.
76. Pradhan, V., Murthy Balijepalli V. S. K., Khaparde, S. A., An effective model for demand response management systems of residential electricity consumers. *IEEE Systems Journal*, 10, 2, 434-445, 2016.
77. Awad, A. S. A., EL-Fouly T. H. M., Salama, M. M. A., Optimal ess allocation for load management application. *IEEE Transactions on Power Systems*, 30, 1, 327-336, 2015.
78. Hauer, A., Quinnell, J et al., Energy storage technologies-characteristics, comparison, and synergies, in: *Transition to Renewable Energy Systems*, 1st ed., pp. 557-577, Wiley-VCH, 2013.
79. Luo, X., Wang, J., Dooner, M., Clarke, J., Overview of current development in electrical energy storage technologies and the application potential in power system operation. *Applied Energy*, 137, 511-536, 2015.
80. Paliwal, P., Patidar, N. P., Nema, R. K., Planning of grid integrated distributed generators: A review of technology, objectives and techniques. *Renewable and Sustainable Energy Reviews*, 40, 557-570, 2014.
81. Electricity Storage Association (ESA), http://www.electricitystorage.org/, 2009.
82. Deane, J., Ó Gallachóir, B. P., McKeogh, E. J., Techno-economic review of existing and new pumped hydro energy storage plant. *Renewable and Sustainable Energy Reviews*, 14, 4, 1293-1302, 2010.
83. Simmons, J., Barnhart, A., Reynolds, S., Young-Jun, S., Study of compressed air energy storage with grid and photovoltaic energy generation. *The Arizona Research Institute for Solar Energy (AzRISE)-APS Final Draft Report, Compressed Air Energy Storage and Photovoltaics Study*, University of Arizona, 2010.
84. Levene, J. I., Mann, M. K., Margolis, R., Milbrandt, A., An analysis of hydrogen production from renewable electricity sources. *Solar Energy*, 81,6, 773-780, 2007.
85. O'Donnell L., Maine, E., Techno-economic analysis of hydrogen production using FBMR technology. *2012 Proceedings of PICMET '12: Technology Management for Emerging Technologies*, Vancouver, BC, 829-835, 2012.

86. Ribeiro, P. F., Johnson, B. K., Crow, M. L., Arsoy, A., Liu, Y., Energy storage systems for advanced power applications," *Proceedings of the IEEE* 89, 12, 1744-1756, 2001.
87. Mahlia, T., Saktisahdan, T., Jannifar, A., Hasan, M., Matseelar, H., A review of available methods and development on energy storage; technology update. *Renewable and Sustainable Energy Reviews*, 33, 532-545, 2014.
88. Luo, F., Meng, K., Dong, Z. Y., Zheng, Y., Chen Y., Wong, K. P.Coordinated operational planning for wind farm with battery energy storage system. *IEEE Transactions on Sustainable Energy*, 6, 1, 253-262,. 2015.
89. Chakraborty, S., Senjyu, T., Toyama, H., Saber, A. Y., Funabashi, T., Determination methodology for optimising the energy storage size for power system. *IET Generation, Transmission & Distribution*, 3, 11, 987-999, 2009.
90. Divya, K., Østergaard, J., Battery energy storage technology for power systems—An overview. *Electric Power Systems Research*, 79, 511-520, 2009.
91. Fang, H., Gong, C., Li, C., Li, X., Su, H., Gu, L., A surrogate model based nested optimization framework for inverse problem considering interval uncertainty. *Structural and Multidisciplinary Optimization*, 58, 3, 869-883, 2018.
92. Xiao, H., Pei, W., Dong, Z., Kong, L., Bi-level planning for integrated energy systems incorporating demand response and energy storage under uncertain environments using novel metamodel. *CSEE Journal of Power and Energy Systems*, 4, 2, 155-167, 2018.
93. Sinha, A., Malo, P., Deb, K., A review on bilevel optimization: From classical to evolutionary approaches and applications. *IEEE Transactions on Evolutionary Computation*, 22, 2, 276-295, 2018.
94. Sharma, S., Niazi, K. R., Verma, K., Thokar, R. A., Bilevel optimization framework for impact analysis of DR on optimal accommodation of PV and BESS in distribution system. *International Transactions on Electrical Energy Systems*, 29, 9, e12062, 2019.
95. Deb, K., Sinha, A., An efficient and accurate solution methodology for bilevel multi objective programming problems using a hybrid evolutionary-local-search algorithm. *Evolutionary Computation*, 18, 3, 403-449, 2010.
96. Yeniay, Ö., Penalty function methods for constrained optimization with genetic algorithms. *Mathematical and Computational Applications*, 10, 1, 45-56, 2005.
97. Joshi, K. A., Pindoriya, N. M., Srivastava, A. K., A two-stage fuzzy multiobjective optimization for phase-sensitive day-ahead dispatch of battery energy storage system. *IEEE Systems Journal*, 12, 4, 3649-3660, 2018.
98. Rui, L., Wei, W., Zhe, C., Xuezhi, W., Optimal planning of energy storage system in active distribution system based on fuzzy multi-objective bi-level optimization. *Journal of Modern Power Systems and Clean Energy*, 6, 2, 342-355, 2018.

99. Jadoun, V. K., Gupta, N., Niazi, K. R., Swarnkar, A., Modulated particle swarm optimization for economic emission dispatch. *International Journal of Electrical Power and Energy Systems*, 73, 80-88, 2015.
100. Savier, J., Das, D., Loss allocation to consumers before and after reconfiguration of radial distribution networks. *International Journal of Electrical Power and Energy Systems*, 33,3, 540-549, 2011.
101. Meena, N. K., Swarnkar, A., Gupta N., Niazi, K. R., Dispatchable solar photovoltaic power generation planning for distribution systems. *2017 IEEE International Conference on Industrial and Information Systems (ICIIS)*, Peradeniya, 1-6, 2017.
102. Odili, J. B., Kahar, M. N. M., Anwar, S., African buffalo optimization: A swarm intelligence technique. *Procedia Computer Science*, 76, 443-448, 2015.
103. Wilson, D. S., Altruism and organism: Disentangling the themes of multilevel selection theory. *The American Naturalist*, 150, S1, S122–S134, 1997.
104. Baran, M. E., Wu, F. F., Network reconfiguration in distribution systems for loss reduction and load balancing. *IEEE Transactions on Power Delivery*, 4, 2,1401-1407, 1989.
105. Thokar, R. A., Gupta, N., Niazi, K. R., Swarnkar, A., Sharma, S., Meena, N. K., Optimal integration and management of solar generation and battery storage system indistribution systems under uncertain environment. *International Journal of Renewable Energy Research*, 10, 1, 1-12, 2020.
106. N. Kanwar, N. Gupta, K. R. Niazi, and A. Swarnkar, "Optimal distributed resource planning for microgrids under uncertain environment," *IET Renewable Power Generation*, vol. 12, no. 2, pp. 244–251, 2018.
107. Rao, R. S., Ravindra, K., Satish, K., Narasimham, S. V. L., Power loss minimization in distribution system using network reconfiguration in the presence of distributed generation. *IEEE Transactions on Power Systems*, 28, 1, 317-325, 2013.

Index

Adoption, 49, 118, 202
Algorithm, 62–65, 73–95, 105, 111, 119, 122, 124, 207, 213, 249, 251, 253, 263, 285–286, 288–292, 302, 308–309, 319
Alkaline, 8, 146, 149
Active distribution network, 199, 243, 248, 250, 253, 274, 302, 316
Air excess ratio, 145, 147, 151, 159, 162, 165
Ancillary services, 22, 44, 63, 71, 201, 202, 237, 283–284
Arbitrage benefits, 283–284, 286–287, 289, 293–295, 301
Artificial intelligence, 64
Asset deferral, 21
Augmented control, 161–162

Barriers, 11, 168–169
Batteries, 6, 13–18, 23, 34, 38, 40–50, 62, 69–71, 73, 79, 81–96, 107, 109, 118, 143, 147, 151, 166–168, 177–179, 181–184, 186–190, 194, 203, 207, 232, 300, 301
Battery, 5, 14–18, 20, 23, 36, 38, 40–42, 46–48, 59, 63, 65, 67, 69–72, 79, 81–83, 85–93, 96, 109, 120, 143, 146, 148, 151, 165, 177–182, 185–186, 188–192, 194, 201, 203, 206–216, 224–225, 228–229, 235, 237, 282–283, 285, 287–289, 294, 298–301, 310
Bromine, 16–18

Battery electric vehicle, 79, 83, 85
Battery energy storage system, 63, 67, 88, 143, 178, 186, 188, 194, 201, 203, 235, 282, 283, 294, 299, 300
Bidirectional power flow, 2, 200
Bilateral transactions, 128, 137
Biogas generator, 188, 189
Black start, 21, 205, 293, 295–296
Blackouts, 106
Buck-boost converter, 71

Cadmium, 15, 38, 89
Capacitors (see: capacitor), 38, 40, 46, 62, 67, 70, 73, 95, 107, 109
Ceramic, 148, 180
Circuit, 46, 109–111, 115, 147, 152, 180
Collide, 37
Congestion, 22, 48, 82, 205, 212, 287, 296
Conservation, 31, 250, 254, 259, 267
Controller, 61, 66, 69, 119–123, 127, 128, 134, 135, 137, 161, 169, 206, 207, 213, 214
Converters, 61, 94, 95, 208
Corrosion, 18, 184
Curtailment, 4, 22, 295
Compressed air energy storage (CAES), 5, 7, 9, 23, 24, 27, 41, 84, 296, 297
Capacitive energy storage, 5, 105, 107–108, 139–140

329

Capacity utilization factor, 189
Capital cost, 20, 23, 24, 36, 49, 88, 185, 187, 189, 192, 223, 224, 298
Ceramic material, 148, 149
Charging time, 15, 79, 143, 146
Chemical energy storage system, 6
Climatic, 34, 46, 79, 80, 85, 168, 282
Combustion, 25, 80, 82, 86, 148
Commodity storage, 21
Contemporary/modern distribution networks, 281, 282
Contract violation, 128–129
Cost of energy, 36, 72, 191, 194, 199, 213, 224, 235, 237, 243, 248, 261, 262, 272, 287, 293

Damping, 60, 61, 137
Database, 203
Day-ahead, 57, 213
DC link capacitor, 67–69, 157
Degradation, 60, 163, 187, 299
Deloading, 62, 66, 67, 69
Dependence, 9, 49, 50
Depicted, 7, 16, 60, 158, 159, 162, 184, 250, 268
Depth of discharge, 8, 20, 113, 214, 215, 299, 301
Depths, 39
Derating factor, 189
Deregulated, 105, 121, 123, 128, 137
Deregulation, 2, 200
Diagnosis, 50
Dielectric, 92, 109
Dimensions, 168, 309
Discharge, 8, 9, 11, 13–16, 20, 23, 24, 27, 28, 47, 68–70, 81, 88, 89, 113, 115, 117, 123, 179, 180, 183, 187, 190, 206, 214, 215, 220, 277, 283, 285, 286, 288, 294, 295, 299–301
Discharge rate, 8, 9, 11, 19, 20, 23, 27, 299
DISCOS, 128–130
Discrete, 286, 309

Dispatchable sources, 284, 286, 293, 311
Disruption, 108, 111, 124
Distributed generation (DG), 3, 42, 48, 169, 200, 206, 213, 235
Distributed renewable energy resources (DRERs), 282, 286
Double layer capacitor (DLC), 12, 83, 94, 107
Droop, 65, 66, 68
Droop control, 66, 68
Durable, 38, 168

Earth, 39
Electro-chemical, 7, 84, 157
Electrodes, 12–15, 18, 38, 90, 108, 109, 147–148, 178, 180, 181, 184, 185
Electrolytes, 8, 15, 16, 182, 183
Emergency, 57, 203–205, 235
Environmental-friendly, 167, 168
Evaluation, 54, 81, 191, 215, 288
Exhaustive, 283, 286, 288, 291
Exploitation, 82, 302, 308, 309
Economic growth, 2
Electric grid, 37, 106
Electric vehicles, 41, 43, 79, 80, 81, 82, 87, 200–202
Electrical energy storage, 13, 38, 105–106, 116, 118
Electrochemical energy storage system, 14
Electrolyte, 14, 48, 149, 183
Electromagnetic energy storage system, 14
Electrostatic energy storage system, 12
End-consumer side, 290, 293, 294, 295
Energy density, 13, 16, 20, 23, 27, 28, 41, 81, 85, 92, 107–109, 113, 117, 178, 181, 182, 187, 299, 300, 301, 310
Energy management, 22, 23, 94, 96, 201, 210, 213–214, 237
Energy storage systems (ESS), 1, 4–9, 12, 19, 21, 28, 35, 41, 44, 46–48,

INDEX 331

59, 62–63, 66, 67, 70, 73, 79, 81, 82, 86, 88, 93, 96, 105–107, 178, 182, 186–188, 194–195, 199, 201–205, 210, 214, 235, 237, 243, 248, 253, 258, 281, 283, 285, 294, 299, 300
Energy Storage Technologies (ESTS), 4, 23, 40, 45, 46, 49, 59, 178, 184, 194, 202, 290, 293, 294, 296, 299, 301, 317
ESS Allocation, 284–292

Failures, 35, 50, 106, 117
Fast-acting, 105, 160
Firming, 22
Fluctuates, 168, 297
Flywheel, 5, 6, 9–11, 26, 30, 53, 57, 67–69, 76, 83, 91, 94, 120, 202, 294, 297, 299
Fossil, 2, 6, 25, 51, 54, 60, 79, 145, 168, 186, 199, 201, 209, 212, 282
Fuzzy-based, 289
Flow battery energy storage system, 16, 194
Flywheel energy storage system, 9, 10, 69, 70, 92, 95
Fuel cell, 6–8, 23, 27, 39, 40, 92, 95, 107, 120, 143, 145–153, 160, 164–170, 202, 294, 299
Fuel utilization factor, 145, 151, 154, 156, 159–162, 169

GENCO, 128–130, 133, 136
Global, 6, 34, 41, 43, 54, 56, 96, 97, 109, 148, 168, 224, 246, 275, 287–89, 291, 292, 308
Greenhouse, 2, 6, 97, 98, 102, 106, 167, 177, 197, 199, 210, 225, 227, 229, 231, 235, 237
Grid-connected, 52, 55, 65, 73, 75, 143, 156, 157, 160, 162, 164, 165, 169, 171, 173, 174, 178, 186, 201, 210, 224, 320

Grids, 21, 41, 50, 57, 70, 72, 101, 172, 195, 240, 254, 321
Global warming, 6, 34, 148
Graphite, 16, 179, 184
Grid angular stability, 21
Grid voltage support, 21

Harmonic, 186
High-cost, 91, 114
Hydroxide, 15, 88, 89, 149, 183
Heuristic methods, 286, 289
HOMER, 192
Hybrid electric vehicle, 81, 82, 86, 87
Hydrogen ESS (HESS), 6

Impedance, 157, 179, 306
Inductance, 12, 115
Inductor, 12, 114, 115
Inertia, 10, 59–74, 76–78, 283
Inertial, 66, 70, 77, 78, 118, 139, 283
Inflammable, 181
Inorganic, 101
Instability, 1, 60, 63, 203
Inverters, 22, 36, 67, 68, 75, 156, 170, 205
Irradiance, 224, 239, 244, 254–258
Irrigation, 67
Islanded, 67
Isothermal, 9
Internal-layer, 304

Justified, 287, 304, 317

Kilogram, 8, 20
Kilowatt-hour, 88

Lead-acid, 15, 30, 34, 41, 47, 83, 87, 90, 94, 98, 109, 178, 182, 186, 194, 196
Lifecycle, 97, 229
Lifespan, 71, 109, 284, 295, 297–300, 307
Lifetime, 20, 23, 71, 72, 107, 177, 179, 189, 191, 216, 238

Lithium-ion, 16, 30, 34, 38, 41–43, 47, 49, 54, 69–71, 74, 89, 93, 101, 102, 103, 145, 178, 195, 196
Loading, 118, 205, 308, 309
Low-temperature, 12, 18
Low-voltage, 104, 171, 240
Load frequency control, 63, 65, 105, 182, 186

Mass, 8, 10, 11, 82, 97, 118, 124
Measures, 44, 74, 94, 95, 201
Methanol, 8, 144, 149, 175
Microgrid, 1, 4, 50, 63, 65, 67, 70, 177, 182, 186–194, 200, 202, 210, 211, 213, 229, 237
Mixed-integer, 211, 286, 287, 292, 302
Molten, 8, 15, 38, 90, 144, 149, 173
Multi-area, 137
Multi-constraint, 286, 302
Multi-objective, 201, 237, 249, 287, 303
Molten carbonate fuel cell (MCFC), 8, 144, 148, 149, 151, 165, 166
Mechanical energy storage system (MESS), 8, 84
Membrane, 8, 17, 18, 38, 90, 146, 149, 180–185
Modern power system, 2–4, 28, 33, 59, 105, 118, 147, 150, 178, 182, 186, 188, 194, 195, 200
Moth flame optimization, 120, 122, 141

Nanoscale, 52, 102
National, 2, 101, 259, 281, 282
Neural, 144, 174
Nowadays, 1, 2, 14, 47, 156, 169, 178, 183, 282
National grid, 2
Natural fuel flow, 145, 161
Nickel hydroxide, 15, 88, 89
Nickel-cadmium battery, 15

Occurrence, 207, 264
Organic, 88, 179
Oscillations, 21, 121, 124, 130
Oxidation, 6, 16, 18, 90, 91, 172, 180–182, 184, 185
Objective function, 121, 124, 217, 260, 288, 303–305
Operating & replacement cost, 193, 194
Operational planning, 260, 284, 285, 300, 308
Optimal allocation, 243, 249, 282, 284, 285, 302, 311, 317
Optimization technique, 63, 119–120, 122, 263, 284, 286
Oxidation reaction, 180, 184

Pattern, 159, 206, 273, 309
Permanent, 68
Permeability, 184
Pollutants, 41, 83, 238
Pollution, 2, 8, 9, 37, 81, 144, 148, 150, 167, 194, 235
Polyhalide, 182
Polymer, 8, 18, 93, 144, 146, 149, 180
Proportional-integral, 121
PAFC, 8, 144, 149, 151, 165, 166
PEMFC, 8, 144, 146, 148, 149, 150, 151, 165, 166
PHES, 5, 23, 296, 297
Photovoltaic, 3, 36, 37, 42, 45, 49, 50, 59, 60, 73, 105, 120, 188, 189, 246, 250, 253, 254, 282
Polysulfide battery, 18
Power conversion unit, 5
Power density, 15, 16, 19, 20, 23, 27, 88, 91, 108, 113, 300, 301
Power system network, 2, 63
Proportional-integral controller, 121
Prosumer, 2
Pumped thermal energy system, 18

Quadratic, 146, 289, 292
Quasi-two-dimensional, 151

Resilience, 232, 235
Resource, 2, 56, 76, 97, 140, 174, 177, 185, 191, 204, 225, 229, 241, 296, 324, 326
Rooftop, 20, 207
Rotor, 39, 62, 70, 99
Round-trip, 181, 190, 295–298, 300, 306
Ramping, 72, 248, 293
Redox flow battery, 182
Reduction reaction, 180,184
Refined fuel, 161
Renewable Energy Resources/Sources (RERS/RES), 3, 59
Response time, 13, 20, 23, 24, 27, 41, 70, 72, 88, 108, 113, 145, 156, 177, 182, 185, 202, 299
Reverse feedback, 295, 307, 315
Round-trip efficiency, 181, 190, 295–298, 300, 306

Scenarios, 49, 161, 243, 245, 257, 258, 260
Self-discharge, 8, 11, 17, 19, 20, 23, 27, 89, 92, 109, 117, 185, 187, 238, 299, 300
Sensible, 6, 18, 86
Shipboard, 57
Short-duration, 23, 114
Silicon, 37, 110, 179
Solid-oxide, 143, 144, 149, 152, 153, 155, 157, 172, 174
Superconductors, 12, 40, 46, 107, 114
Simultaneous allocation, 302, 310
Smart distribution networks (SDNs), 282
SMES, 5, 7, 24, 26, 46, 105, 107, 190, 113–120, 124, 130, 136, 137, 297, 299
Sodium-sulfur battery, 15
Solar Photovoltaic (SPV), 3, 37, 182, 189, 282
Solid-oxide fuel cells, 150–152

Stability, 106–107, 130, 136, 139
State-of-charge (SOC), 20, 284, 305, 315
Stochastic variable module, 256–257

TCPS, 119, 120
Techno-economic, 29, 57, 143, 166, 167, 174, 177, 196, 197, 199, 201, 213, 224, 225, 235, 237, 248, 314, 324
Thermal-hydro-gas, 121, 122
Tie-line, 121–124, 129–131, 134, 139
Time-span, 96
Thermal energy storage system, 18
Traditional, 105, 120–122, 124, 137, 141
Transfer function, 112, 115, 156
Transmission curtailment reduction, 22
Two-layer optimization +, 302
Two-stage coordinated optimization model, 259–260

Ultra-capacitor, 7, 12, 70, 91, 107, 120
Utilities, 235, 248, 284

Vacuum, 11, 12, 39, 91
Vanadium, 16, 31, 87, 89, 194, 196, 197, 198, 300
Volts, 154, 164
Vulnerability, 170
Vanadium redox battery, 16
Virtual inertia, 59, 61–62, 66–69, 71
Voltage regulation, 63, 69, 70, 243, 274, 282, 296

Watt-hour, 20
Weather, 37, 200
Worldwide, 43, 59, 108, 143, 146

Yearly, 191, 261, 262, 264

Zinc-bromine, 16, 17

Also of Interest

Check out these other related titles from Scrivener Publishing

Energy Storage, edited by Umakanta Sahoo, ISBN 9781119555513. Written and edited by a team of well-known and respected experts in the field, this new volume on energy storage presents the state-of-the-art developments and challenges in the field of renewable energy systems for sustainability and scalability for engineers, researchers, academicians, industry professionals, consultants, and designers. NOW AVAILABLE!

Energy Storage 2nd Edition, by Ralph Zito and Haleh Ardibili, ISBN 9781119083597. A revision of the groundbreaking study of methods for storing energy on a massive scale to be used in wind, solar, and other renewable energy systems. NOW AVAILABLE!

Hybrid Renewable Energy Systems, edited by Umakanta Sahoo, ISBN 9781119555575. Edited and written by some of the world's top experts in renewable energy, this is the most comprehensive and in-depth volume on hybrid renewable energy systems available, a must-have for any engineer, scientist, or student. NOW AVAILABLE!

Progress in Solar Energy Technology and Applications, edited by Umakanta Sahoo, ISBN 9781119555605. This first volume in the new groundbreaking series, Advances in Renewable Energy, covers the latest concepts, trends, techniques, processes, and materials in solar energy, focusing on the state-of-the-art for the field and written by a group of world-renowned experts. NOW AVAILABLE!

A Polygeneration Process Concept for Hybrid Solar and Biomass Power Plants: Simulation, Modeling, and Optimization, by Umakanta Sahoo, ISBN 9781119536093. This is the most comprehensive and in-depth study of the theory and practical applications of a new and groundbreaking method for the energy industry to "go green" with renewable and alternative energy sources. NOW AVAILABLE!

Nuclear Power: Policies, Practices, and the Future, by Darryl Siemer, ISBN 9781119657781. Written from an engineer's perspective, this is a treatise on the state of nuclear power today, its benefits, and its future, focusing on both policy and technological issues. *NOW AVAILABLE!*

Zero-Waste Engineering 2nd Edition: A New Era of Sustainable Technology Development, by M. M. Kahn and M. R. Islam, ISBN 9781119184898. This book outlines how to develop zero-waste engineering following natural pathways that are truly sustainable using methods that have been developed for sustainability, such as solar air conditioning, natural desalination, green building, chemical-free biofuel, fuel cells, scientifically renewable energy, and new mathematical and economic models. *NOW AVAILABLE!*

Sustainable Energy Pricing, by Gary Zatzman, ISBN 9780470901632. In this controversial new volume, the author explores a new science of energy pricing and how it can be done in a way that is sustainable for the world's economy and environment. *NOW AVAILABLE!*

Sustainable Resource Development, by Gary Zatzman, ISBN 9781118290392. Taking a new, fresh look at how the energy industry and we, as a planet, are developing our energy resources, this book looks at what is right and wrong about energy resource development. This book aids engineers and scientists in achieving a true sustainability in this field, both from an economic and environmental perspective. *NOW AVAILABLE!*

The *Greening of Petroleum Operations*, by M. R. Islam *et al.*, ISBN 9780470625903. The state of the art in petroleum operations, from a "green" perspective. *NOW AVAILABLE!*

Emergency Response Management for Offshore Oil Spills, by Nicholas P. Cheremisinoff, PhD, and Anton Davletshin, ISBN 9780470927120. The first book to examine the Deepwater Horizon disaster and offer processes for safety and environmental protection. *NOW AVAILABLE!*

Biogas Production, Edited by Ackmez Mudhoo, ISBN 9781118062852. This volume covers the most cutting-edge pretreatment processes being used and studied today for the production of biogas during anaerobic digestion processes using different feedstocks, in the most efficient and economical methods possible. *NOW AVAILABLE!*

Bioremediation and Sustainability: Research and Applications, Edited by Romeela Mohee and Ackmez Mudhoo, ISBN 9781118062845. Bioremediation and Sustainability is an up-to-date and comprehensive treatment of research and applications for some of the most important low-cost, "green," emerging technologies in chemical and environmental engineering. *NOW AVAILABLE!*

Green Chemistry and Environmental Remediation, Edited by Rashmi Sanghi and Vandana Singh, ISBN 9780470943083. Presents high quality research papers as well as in depth review articles on the new emerging green face of multidimensional environmental chemistry. *NOW AVAILABLE!*

Bioremediation of Petroleum and Petroleum Products, by James Speight and Karuna Arjoon, ISBN 9780470938492. With petroleum-related spills, explosions, and health issues in the headlines almost every day, the issue of remediation of petroleum and petroleum products is taking on increasing importance, for the survival of our environment, our planet, and our future. This book is the first of its kind to explore this difficult issue from an engineering and scientific point of view and offer solutions and reasonable courses of action. *NOW AVAILABLE!*

Printed and bound by CPI Group (UK) Ltd, Croydon, CR0 4YY